Statistical Techniques in Bioassay

Z. Govindarajulu

Statistical Techniques in Bioassay

4 figures and 33 tables, 1988

Basel · München · Paris · London · New York · New Delhi · Singapore · Tokyo · Sydney

Zakkula Govindarajulu

PhD, Professor, University of Kentucky, Department of Statistics, Lexington, Ky.

Fellow of the Institute of Mathematical Statistics, American Statistical Association, Royal Statistical Society (England), and the American Association for the Advancement of Science; Member of the Bernoulli Society; Member of the International Statistical Institute and a member of the National Academy of Sciences (India).

Library of Congress Cataloging-in-Publication Data
　　Govindarajulu, Z.
　　Statistical techniques in bioassay.
　　Includes index.
　　1. Biological assay – Statistical methods. I. Title.
　　[DNLM: 1. Biological Assay – methods. 2. Statistics. QV 771 G721s]
　　QH323.5.G68 1988　574.19′285′072　88-546
　　ISBN 3–8055–4630–0

All rights reserved.
　　No part of this publication may be translated into other languages, reproduced or utilized in any form or by any means, electronic or mechanical, including photocopying, recording, microcopying, or by any information storage and retrieval system, without permission in writing from the publisher.

©　　Copyright 1988 by S. Karger AG, P.O. Box, CH–4009 Basel (Switzerland)
　　Typeset in Hong Kong by Asco Trade Typesetting Limited
　　Printed in Switzerland by Thür AG Offsetdruck, Pratteln
　　ISBN 3–8055–4630–0

*Dedicated to the memory of my parents-in-law,
Dr. and Mrs. Mahanand Gupta*

Contents

Foreword by Byron Wm. Brown, Jr. X
Preface XI

1 Introduction 1
1.1. History of Bioassay 1
1.2. Components of Bioassay 1
1.3. Role of Statistics in Bioassay 2

2 Preliminaries 3
2.1. Types of Biological Assays 3
2.2. Direct Assays 3
2.3. Ratio Estimates 4
2.4. Asymptotic Distribution of Ratio Estimators $\sqrt{n_2}\left(\dfrac{\bar{Y}}{\bar{X}} - \dfrac{\nu}{\mu}\right)$ 5
2.5. Fieller's Theorem 5
2.6. Behrens' Distribution 6
2.7. Confidence Intervals for the Natriuretic Factor Assay 7
2.8. Use of Covariates 10
Appendix: Behrens-Fisher-Sukhatme Distribution 11

3 Algebraic Dose-Response Relationships 12
3.1. Indirect Assays 12
3.2. Dose-Response Relationship in Terms of Regression 12
3.3. Preliminary Guess of the Regression Function 13
3.4. Transformation Leading to Linear Relationships 14
3.5. Nonlinear Regression 15
3.6. Heterogeneity of Variance 17
3.7. Maximum Likelihood Estimates of Parameters 19
3.8. Maximum Likelihood: Iterative Scheme 20
3.9. Estimation of the Relationship for the Standard Preparation 23
3.10. Estimation of the Slope 24
3.11. Estimation Based on Simultaneous Tests 25
Appendix: Regression Models 25

4 The Logit Approach 28
4.1. Introduction 28
4.2. Case when the Dose-Response Curve for the Standard Preparation Is Unknown .. 30
4.3. Quantal Responses 32
4.4. Linear Transformations for Sigmoid Curves: Tolerance Distribution 32

4.5.	Importance and Properties of the Logistic Curve	34
4.6.	Estimation of the Parameters	35
4.7.	Estimations of the Parameters in the Probit by the Method of Maximum Likelihood	38
4.8.	Other Available Methods	40
4.9.	Method of Minimum Logit χ^2	41
4.10.	Goodness-of-Fit Tests	43
4.11.	Spearman-Karber Estimator	44
4.12.	Reed-Muench Estimate	51
4.13.	Dragstadt-Behrens Estimator	52
4.14.	Application of the Spearman Technique to the Estimation of the Density of Organisms	52
4.15.	Quantit Analysis (Refinement of the Quantal Assay)	55
4.16.	Planning a Quantal Assay	58
4.17.	Dose Allocation Schemes in Logit Analysis	59
	Appendix: Justification for Anscombe's Correction for the Logits l_i	63

5 Other Methods of Estimating the Parameters ... 65

5.1.	Case of Two Dose Levels	65
5.2.	Natural Mortality	67
5.3.	Estimation of Relative Potency	71
5.4.	An Optimal Property of the Logit and Probit Estimates	73
5.5.	Confidence Bands for the Logistic Response Curve	77
5.6.	Weighted Least Squares Approach	79

6 The Angular Response Curve ... 81

6.1.	Estimation by the Method of Maximum Likelihood	81
6.2.	Alternative Method of Estimation	83
6.3.	Comparison of Various Methods	83
6.4.	Other Models	84
6.5.	Comparison of Maximum Likelihood and Minimum χ^2 Estimates	84
6.6.	More on Probit Analysis	85
6.7.	Linear Logistic Model in 2×2 Contingency Tables	88
6.8.	Comparison of Several 2×2 Contingency Models	91

7 Estimation of Points on the Quantal Response Function ... 94

7.1.	Robbins-Monro Process	94
7.2.	Robbins and Monro Procedure	96
7.3.	Parametric Estimation	100
7.4.	Up and Down Rule	103

8 Sequential Up and Down Methods ... 107

8.1.	Up and Down Transformed Response Rule	107
8.2.	Stopping Rules	108
8.3.	Up and Down Method	108
8.4.	Finite Markov Chain Approach	110
8.5.	Estimation of the Slope	111
8.6.	Expected Values of the Sample Size	111
8.7.	Up and Down Methods with Multiple Sampling	111
8.8.	Estimation of Extreme Quantiles	113
	Appendix: Limiting Distribution of $D = (N - \frac{1}{2})\Delta$	115

9   *Estimation of 'Safe Doses'* 117

9.1.   Models for Carcinogenic Rates 117
9.2.   Maximum Likelihood Estimation of the Parameters 118
9.3.   Convex Programming Algorithms 120
9.4.   Point Estimation and Confidence Intervals for 'Safe Doses' 121
9.5.   The Mantel-Bryan Model 123
9.6.   Dose-Response Relationships Based on Dichotomous Data 124
9.7.   Optimal Designs in Carcinogen Experiments 127

10   *Bayesian Bioassay* 133

10.1.   Introduction 133
10.2.   The Dirichlet Prior 134
10.3.   The Bayes Solution for Squared Error Loss 135
10.4.   The Alternate Bayes Approaches 136
10.5.   Bayes Binomial Estimators 139
10.6.   Selecting the Prior Sample Size 140
10.7.   Adaptive Estimators 140
10.8.   Mean Square Error Comparisons 142
10.9.   Bayes Estimate of the Median Effective Dose 143

11   *Radioimmunoassays* 145

11.1.   Introduction 145
11.2.   Isotope Displacement Immunoassay 145
11.3.   Analysis of Variance of the McHugh and Meinert Model 149
11.4.   Other Models for Radioimmunoassays 150
11.5.   Assay Quality Control 151
11.6.   Counting Statistics 152
11.7.   Calibration Curve Fitting 152
11.8.   Principles of Curve-Fitting 155

References 156
Subject Index 163

Foreword

I am happy indeed to write this foreword. Raju and I were two of Richard Savage's first thesis students at the University of Minnesota. My thesis was on certain aspects of quantal assay. Raju was interested and helpful. We have kept in touch and I am gratified to see this product of an interest partially fostered by me then and encouraged in the succeeding years.

This is a book that has been needed for a long time. Every biostatistical consultant is faced, now and again, with a question on the design or analysis of a quantal assay. Quantal assays will always be an important tool in experimental biomedicine. This book presents the statistical aspects of quantal assay and analysis in a mathematically concise, rigorous and modern format. Coverage includes some material found at the present only in periodicals and technical reports. It will be a useful addition to the armamentarium of the biostatistical consultant, whether fledgling or veteran. I believe the book will also serve well as a text for a course that focuses on quantal assay, for students of statistics interested in biomedical consulting, or as a supplementary text for a more generally applied statistics course for such students.

Byron Wm. Brown, Jr.
Professor and Head
Division of Biostatistics,
Department of Family, Community and Preventive Medicine,
Stanford University School of Medicine, Palo Alto, Calif.

Preface

Since the beginning of the 20th century, there has been a lot of activity in developing statistical methods for analyzing biological data. The development of the probit method is originally due to Gaddum and Bliss. The two important methods of analyzing biological data are: (1) the probit method, and (2) the logit method. Finney [1971] has written an exhaustive treatise on the probit method; Ashton [1972] a short monograph on the logit approach. Here we give equal importance to both the probit and the logit approaches and deal with other approaches to bioassay. Chapters first deal with direct and indirect assays. The logit approach is then covered. Further chapters focus on the angular response curve and other methods, while other chapters consider sequential methods. Readers will also find chapters devoted to estimation of low doses, Bayesian methods, and radioimmunoassays. There have been recent developments, especially in robust estimation methods in bioassay; however, these are not included in this book. More than 200 references are cited and given in a list at the end; this list is by no means complete. A basic course in statistical inference is all that is required of the readers.

I shall appreciate readers drawing my attention to any shortcomings or errors found in this book. This book grew out of my lecture notes based on a course in bioassay given at the University of Kentucky during several summers. A quarter or semester's course on bioassay can be taught out of this book. Selection of the appropriate chapters depends upon the emphasis of the course and the interests of the audience.

I give special thanks to my students who were the involuntary 'guinea pigs' in the course I taught. I am thankful to Vicki Kenney, Debra Arterburn, Brian Moses and Susan Hamilton for the excellent typing of the manuscript. I thank the Department of Statistics for its support and other help. It is also a pleasure to thank the staff of S. Karger AG for generous help and excellent cooperation throughout this project. I thank Professors Byron Brown of Stanford University, Charles Bell of San Diego State University and Bartholomew Hsi of the University of Texas at Houston for reading the manuscript in its early stages and making very helpful comments and suggestions. For generous permission to reproduce tables and/or to use material, my special thanks go to Dr. Margaret Wesley, American Association for the Advancement of Science, Association of Applied Biologists, the American Statistical Association, the Biometric Society, the Biometrika Trustees, the Biochemical Society, the Institute of Mathematical Statistics, the Royal Statistical Society, the Society for Industrial and Applied Mathematics, Charles Griffin Publishers, the Methuen Company, Cambridge University Press, the MacMillan Publishing Company, Freeman and Company, the Longman House (United Kingdom), Marcel Dekker Inc., MIT Press, and the University of California Press.

Z. Govindarajulu

'Sound and sufficient reason falls, after all, to the share of but few men,
and those few men exert their influence in silence.'

Johann Wolfgang von Goethe

'Every man should use his intellect, not as he uses his lamp in the study,
only for his own seeing,
but as the lighthouse uses its lamps,
that those afar off on the sea may see the shining, and learn their way.'

Henry Ward Beecher

1 Introduction

Definition

Biological assays (bioassay, for short) are methods for estimating the potency of a drug or material by utilizing the reaction caused by its application to experimental subjects that are living. For example, how do pharmacists know that 6 aspirin tablets can be fatal to a child? Qualitative assessments of material do not pose any great problems. Quantitative assays of the material are our main concern.

Examples

(1) A study of the effects of different samples of insulin on the blood sugar of rabbits is not necessarily a bioassay problem. However, it will be if the investigator is interested in the estimation of the potencies of the samples on a scale of standard units of insulin.
(2) A study of the responses of potatoes to various phosphate fertilizers would not be a bioassay; it is a bioassay if one is interested in using the yields of potatoes in assessing the potency of a natural rock phosphate. First, there is a physical quantity (like the weight of an animal) which we call factor X whose levels or doses can be controlled. The effect of the factor X will be called the response. The relation between the dose and the response will be described by means of a graph or an algebraic equation.

1.1. History of Bioassay

Although bioassay is regarded as a recent development, the essence of modern quantal response techniques were used by people in early times. Emmens [1948] was the first to consider the statistical aspects of bioassay. Coward [1947] and Gaddum [1948] considered the biological aspects of the assay.

1.2. Components of Bioassay

The typical bioassay involves a stimulus (for example, a vitamin or a drug) applied to a subject (for example, an animal, a piece of animal tissue, etc). The level of the stimulus can be varied and the effect of the stimulus on the subject can be measured in terms of a characteristic which will be called response. Although a relationship between stimulus and response (which can be characterized by means of an algebraic expression) might exist, the response is subjected to a random error. The relationship can be used to study the potency of a dose from the response it produces.

The estimate of potency is always relative to a standard preparation of the stimulus, which may be a convenient working standard adopted in a laboratory. A *test preparation* of the stimulus, having an unknown potency, is assayed to find the mean response to a selected drug. Next we find the dose of the standard preparation which produces the same mean response. The ratio of the two equally effective doses is an estimate of the potency of the test preparation relative to that of the standard.

We think of an ideal situation where the test and standard preparations are identical in their biologically active ingredient and differ only in the degree of dilution by inactive materials (such as solvents) to which they are subjected. The question is whether a relative potency estimated from one response or one species of subjects can be assumed to have even approximate validity for another response or species.

1.3. *Role of Statistics in Bioassay*

A statistician can make the following contributions: (1) advise on the general statistical principles underlying the assay method, (2) devise a good experimental design that gives the most useful and reliable results, and (3) analyze the data making use of all the evidence on potency. Here design consists of choosing the number of levels of doses at which each preparation is to be tested, the number of subjects to be used at each dose, the method of allocating subjects to doses, the order in which subjects under each dose should be treated and measured, and other aspects of the experiment. Needless to emphasize that a sound design is as important as the statistical analysis of data.

2 Preliminaries[1]

2.1. *Types of Biological Assays*

Besides the purely qualitative assays, there are three main types of biological assays that are commonly used for numerical evaluation of potencies: (1) direct assays; (2) indirect assays based on quantitative responses, and (3) indirect assays based on quantal ('all or nothing') responses. Assays (2) and (3) are similar with respect to their statistical analysis. Both make use of dose-response regression relationships.

2.2. *Direct Assays*

Direct assays are of fundamental importance. A direct assay is based on the principle that doses of the standard and test preparations, sufficient to produce a specified response, are directly measured. The ratio of these two estimates the potency of the test preparation relative to the standard.

Definition

The potency is the amount of the standard equivalent in effect to one unit of the test preparation. An example of a direct assay is the 'cat' method for the assay of digitalis. The standard or the test preparation is infused into the blood stream of a cat until the heart stops beating. The dose is equal to the total period of infusion multiplied by the rate. This experiment is replicated on several cats for each preparation and the average doses are obtained.

Example

Assay of Atrial Natriuretic Factor. The following table gives sodium excretions (μg/min) by male Sprague-Dawley rats. The atrial extract supernatant was injected intravenously (0.2 ml over 30–45 s). Sodium excretion rates for 15 min during the administration of the extracts are shown in table 2.1.

---

[1] Finney [1971b] served as a source for the material of this chapter.

Table 2.1. Excretion by Sprague-Dawley rats [Chimoskey et al., 1984]

	Normal atrial extract	B10 14.6 atrial extract
	1.061	0.406
	2.176	1.490
	2.580	0.594
	7.012	1.064
	2.296	1.332
	3.743	4.172
	4.547	0.917
Mean ± SEM	3.345 ± 0.745	1.425 ± 0.480

The potency of B10 14.6 atrial extract relative to the normal one is $R_B = 1.425/3.345 = 0.426$. That is, one unit of B10 14.6 extract is equivalent to 0.426 units of the normal extract.

2.3. Ratio Estimates

Let

$$X_i = \mu + \varepsilon_i (i = 1,\ldots,n_1) \quad \text{and} \quad Y_j = \nu + \eta_j (j = 1,\ldots,n_2).$$

Then

$\bar{X} = \mu + \bar{\varepsilon}, \bar{Y} = \nu + \bar{\eta}$ and

$$\frac{\bar{Y}}{\bar{X}} = \frac{\nu + \bar{\eta}}{\mu + \bar{\varepsilon}} = \frac{\nu}{\mu}\left(\frac{1 + \frac{\bar{\eta}}{\nu}}{1 + \frac{\bar{\varepsilon}}{\mu}}\right)$$

$$= \frac{\nu}{\mu}\left(1 + \frac{\bar{\eta}}{\nu}\right)\left(1 - \frac{\bar{\varepsilon}}{\mu} + \frac{\bar{\varepsilon}^2}{\mu^2} - \cdots\right)$$

$$= \frac{\nu}{\mu}\left(1 + \frac{\bar{\eta}}{\nu} - \frac{\bar{\varepsilon}}{\mu} - \frac{\bar{\eta}\bar{\varepsilon}}{\mu\nu} + \frac{\bar{\varepsilon}^2}{\mu^2} - \cdots\right).$$

Hence

$$E(\bar{Y}/\bar{X}) = \frac{\nu}{\mu}\left[1 + \frac{\sigma_1^2}{n_1 \mu^2} - \text{cov}\frac{(\bar{\varepsilon},\bar{\eta})}{\mu\nu} + \cdots\right]$$

$$\doteq \frac{\nu}{\mu}\left(1 + \frac{\sigma_1^2}{n_1 \mu^2}\right) \text{ if X and Y are independent,}$$

$$\operatorname{var}(\bar{Y}/\bar{X}) = \left(\frac{v}{\mu}\right)^2 \left[\operatorname{var}\left(\frac{\bar{\eta}}{v}\right) + \operatorname{var}\left(\frac{\bar{\varepsilon}}{\mu}\right) - 2\operatorname{cov}\frac{(\bar{\varepsilon},\bar{\eta})}{\mu v} + \cdots\right]$$

$$= \left(\frac{v}{\mu}\right)^2 \left[\frac{\sigma_2^2}{n_2 v^2} + \frac{\sigma_1^2}{n_1 \mu^2}\right]$$

$$\doteq \sigma^2 \left[\frac{1}{n_2 \mu^2} + \left(\frac{v}{\mu}\right)^2 \frac{1}{n_1 \mu^2}\right],$$

provided X and Y are independent and have a common variance σ^2. (Notice that we assume that $\left|\frac{\bar{\varepsilon}}{\mu}\right| < 1$.)

2.4. *Asymptotic Distribution of Ratio Estimators* $\sqrt{n_2}\left(\frac{\bar{Y}}{\bar{X}} - \frac{v}{\mu}\right)$

$$\frac{\bar{Y}}{\bar{X}} - \frac{v}{\mu} = \frac{\bar{Y}\mu - \bar{X}v}{\bar{X}\mu} = \frac{\mu(\bar{Y} - v) - v(\bar{X} - \mu)}{\bar{X}\mu}$$

$$n_2^{1/2}\left(\frac{\bar{Y}}{\bar{X}} - \frac{v}{\mu}\right) = n_2^{1/2}\left(\frac{\bar{Y} - v}{\bar{X}}\right) - n_2^{1/2}\frac{v}{\mu}\left(\frac{\bar{X} - \mu}{\bar{X}}\right)$$

$$\approx n_2^{1/2}\left(\frac{\bar{Y} - v}{\mu}\right) - n_2^{1/2}\frac{v}{\mu^2}(\bar{X} - \mu).$$

Hence

$$n_2^{1/2}\left(\frac{\bar{Y}}{\bar{X}} - \frac{v}{\mu}\right)$$

is asymptotically normal with mean 0 and variance

$$\frac{\sigma_2^2}{\mu^2} + \frac{n_2}{n_1}\frac{v^2}{\mu^4}\sigma_1^2,$$

if X and Y are independent. If $\sigma_1^2 = \sigma_2^2$, then the asymptotic variance is

$$\frac{\sigma^2}{\mu^2}\left[1 + \frac{n_2}{n_1}\frac{v^2}{\mu^2}\right].$$

2.5. *Fieller's Theorem* [Fieller, 1940, 1944]

Theorem

Let μ, v be two unknown parameters and let $\rho = \mu/v$. Let a and b be unbiased estimators for μ and v, respectively. Assume that a and b are linear in observations that are normally distributed.

Let $R = a/b$ be an estimate of ρ. Then the upper and lower confidence limits for ρ are

$$R_L, R_U = \left\{ R - \frac{gv_{12}}{v_{22}} \pm \frac{ts}{b}\left[v_{11} - 2Rv_{12} + R^2 v_{22} - g\left(v_{11} - \frac{v_{12}^2}{v_{22}}\right)\right]^{1/2}\right\} \bigg/ (1 - g)$$

where $g = t^2 s^2 v_{22}/b^2$ and s^2 is an error mean square having m degrees of freedom and the estimates of the variances and covariance of a and b are $v_{11}s^2$, $v_{22}s^2$ and $s^2 v_{12}$, respectively; also $t = t_{m, 1-\alpha/2}$.

Proof. Consider $a - \rho b$. Then $E(a - \rho b) = 0$ and it has an estimated variance given by

$$s^2(v_{11} - 2\rho v_{12} + \rho^2 v_{22})$$

with m degrees of freedom. Thus,

$$P[(a - \rho b)^2 \leq t^2 s^2 (v_{11} - 2\rho v_{12} + \rho^2 v_{22})] = 1 - \alpha.$$

The two roots for ρ obtained by solving the quadratic in ρ within the probability statement will yield the R_L and R_U given above.

This result will be used in setting confidence limits for the ratio of two means, two regression coefficients or a horizontal distance between two regression lines (because the latter can be expressed algebraically in terms of the ratio of a difference of two means to a regression coefficient.

When b is large in comparison to its standard error, then we can approximately set $g = 0$ and obtain

$$R_L, R_U = R \pm \frac{ts}{b}(v_{11} - 2Rv_{12} + R^2 v_{22})^{1/2}$$

Also, often we will have $v_{12} = 0$.

The general formula is not valid when g exceeds 1.0.

Creasy [1954] and Fieller [1954] propose a solution to the problem of setting confidence intervals for the ratio of normal means using a fiducial approach. Fisher [1956] showed the simple relationship between the solutions of Creasy [1954] and Fieller [1954]. Fieller's confidence intervals are always wider and thus are more conservative than the intervals based on Creasy's fiducial distribution. However, James et al. [1974] point out that Fieller's intervals are not conservative enough. The latter authors provide an interval estimation procedure which is similar to that of Creasy [1954] which is always wider than the Fieller confidence interval (see table 1 of James et al. [1974, p. 181]).

2.6. *Behrens' Distribution*

Let a and b be independent and unbiased estimates of μ and ν and be linear combinations of observations that are normally distributed. Also assume that estimates of var a and var b are given by

$$\widehat{\mathrm{var}\, a} = s_1^2 v_{11}, \quad \widehat{\mathrm{var}\, b} = s_2^2 v_{22}$$

where s_1^2 and s_2^2 are independent mean squares with m_1, m_2 degrees of freedom, respectively. Then the estimated variance of $a - b$ is

$$s_{a-b}^2 = s_1^2 v_{11} + s_2^2 v_{22}.$$

Let

$$D = \frac{(a-b)-(\mu-\nu)}{(s_1^2 v_{11} + s_2^2 v_{22})^{1/2}}.$$

Then D has the Behrens' distribution (D is also called Sukhatme's D-statistic) which is tabulated by Finney [1971b, Appendix III]. Its distribution is defined in terms of the degrees of freedom m_1 and m_2 and the angle θ where

$$\tan\theta = (s_1^2 v_{11}/s_2^2 v_{22})^{1/2}.$$

When $\theta = 0°$, D has t-distribution with m_2 d.f.; when $\theta = 90°$, then D has t-distribution with m_1 d.f. For $0 < \theta < \pi/2$, the distribution of D is between t_{m_1} and t_{m_2}. When $m_1 = m_2$, the value of D for any probability level is about equal, but a little less than the corresponding t-value for all θ. The D-test provides a test of significance for the difference between two means, or two regression coefficients whose variances are based on independent mean squares. For moderately large m_1, m_2, D tends to the standard normal variable in distribution. Next, let a and b denote independent estimates of $\mu = \nu$. Then we consider the weighted estimate given by

$$\bar{a} = [a(s_1^2 v_{11})^{-1} + b(s_2^2 v_{22})^{-1}][(s_1^2 v_{11})^{-1} + (s_2^2 v_{22})^{-1}]^{-1} \text{ and}$$

$$s_{\bar{a}}^2 = [(s_1^2 v_{11})^{-1} + (s_2^2 v_{22})^{-1}]^{-1}.$$

Yates [1939] and Finney [1951] point out that

$$(\bar{a} - \mu)[(s_1^2 v_{11})^{-1} + (s_2^2 v_{22})^{-1}]^{-1/2}$$

follows the Behrens' distribution with d.f. m_1 and m_2 and an angle θ defined by

$$\tan\theta = (s_2^2 v_{22}/s_1^2 v_{11})^{1/2}.$$

Hence, one can test the significance of $\bar{a}$ from a theoretical value μ and also set up a confidence interval for μ as

$$\bar{a} \pm d_{1-\alpha/2}[(s_1^2 v_{11})^{-1} + (s_2^2 v_{22})^{-1}]^{-1/2}$$

where $d_{1-\alpha/2}$ denotes the tabulated value for the probability $1 - \alpha/2$. For incorporating the information about the magnitude of $a - b$ on the distribution of $\bar{a}$, the reader is referred to Fisher [1961a, b].

2.7. Confidence Intervals for the Natriuretic Factor Assay

Let us assume that the sodium excretion rates are normally distributed with constant variance. The estimate of the common variance is

$$s^2 = 7\{(0.745)^2 + (0.480)^2\}/2 = 2.749$$

and the pooled sample standard deviation $s = 1.658$. Here we apply Fieller's theorem (theorem 1) to set up confidence limits for $\rho_B = \mu/\nu$. Let $R_B = a/b$. Then, the estimates of the variances and covariances of a and b, respectively, are

$$v_{11}s^2 = v_{22}s^2 = 2.749/7 = 0.393 \quad \text{and} \quad v_{12} = 0.$$

Thus

if $1 - \alpha = 0.90$, then $t_{12, 0.95} = 1.78$,

$g = (1.78)^2 (0.393)/(3.345)^2 = 0.11$,

$$R_L, R_U = \left\{ 0.426 \pm \frac{(1.78)(1.658)}{\sqrt{7(3.345)}} [1 + (0.426)^2 - (0.111)]^{1/2} \right\} \Big/ (1 - 0.111)$$

$$= \left\{ 0.426 \pm \frac{(1.78)(1.658)}{8.85} (1.034) \right\} \Big/ 0.889$$

$$= \{ 0.426 \pm 0.345 \}/0.889$$

$$= (0.091, 0.867).$$

Remark. Notice that we assumed that the population variances for the two extracts are the same. If they are not the same, then one can apply the improvisation of Fieller's theorem obtained by Finney [1971b, section 2.7].

2.7.1. Dilution Assays

In many assays, the test preparation looks like a dilution (or concentration) of the standard preparation.

An analytical dilution assay is one in which an analysis is made of the effective constituent of a preparation against a standard preparation which has in common all the constituents of the preparation except the one which has an effect on the response of the subjects.

A comparative dilution assay is one in which the two preparations may look alike qualitatively, although they are not the same. For instance, two insecticides of related but different chemical compositions in their effects on an insect species may behave as though one was a dilution of the other. For preparations A and B, let X_A and X_B denote equivalent doses and that B behave like a dilution of A by a factor ρ, the relative potency. Hence

$\rho X_A = X_B$, then

$\rho^2 \operatorname{var} X_A = \operatorname{var} X_B$.

Unless ρ is close to unity, we cannot assume homogeneity of variances. However, by taking logarithmic values we have

$\ln X_A + \ln \rho = \ln X_B$.

We can preserve the homogeneity of variances for log doses. Thus, it seems reasonable to assume normality for the distribution of log dose, since it can vary from $-\infty$ to ∞. The advantages of doing analysis on log dosages are:

(1) All variance estimates can be pooled, which enables one to have more precise estimation.

(2) Since the estimate of the relative potency is obtained as the antilog of the difference of two means rather than the ratio of two means, confidence intervals can be calculated from simple standard error formulae without the use of Fieller's theorem.

Example

Natriuretic factor assay with log sodium excretion rates:

Table 2.2. Sodium excretion (logarithmic values)

	Normal atrial extract	B10 14.6 atrial extract
	0.059	−0.901
	0.777	0.339
	0.948	−0.521
	1.948	0.062
	0.831	0.287
	1.320	1.428
	1.514	−0.087
ΣX_i	7.397	0.667
$\bar{X}$	1.057	0.095
$\Sigma(X_i - \bar{X})^2$	2.209	3.369

One can perform an F test for the hypothesis of equality of variances:

$$F = \frac{3.369/6}{2.209/6} = 1.525$$

which is not significant. Thus the pooled estimate of the variances is

$s^2 = (2.209 + 3.369)/12 = 0.465$ and

$s = 0.682$.

Let $M_B = \ln R_B = \bar{X}_B - \bar{X}_A = 0.095 - 1.057 = -0.962$, where A denotes the normal extract.

$$s^2_{M_B} = s^2\left(\frac{1}{N_B} + \frac{1}{N_A}\right) = 0.465\left(\frac{1}{7} + \frac{1}{7}\right) = (0.364)^2.$$

Hence, assuming that $(M_B - \ln \rho_B)/s$ is distributed as Student's t with 12 degrees of freedom, we obtain for the 90% confidence interval

$M_L = -0.962 - (1.78)(0.364) = -1.610$

$M_U = -0.962 + (1.78)(0.364) = -0.314$

Hence

$R_B = 0.382, \quad R_{L,B} = \exp(M_L) = 0.200, \quad R_{U,B} = 0.731.$

2.7.2. *Precision of Estimates*

Consider the natriuretic factor data (without the ln transformation). We can assume that A and B have the same variability (since the F statistic is 2.41 which is not significant). Then the pooled variance is

$s^2 = 2.749$

$\text{var}(\bar{X}_A) = \text{var}(\bar{X}_B) = s^2/7 = 2.749/7 = 0.393.$

Hence

$$\text{estimate of var}(R_B) = \frac{s^2}{\bar{X}_A^2}\left[1 + \frac{R^2}{7}\right] = \frac{2.749}{(3.345)^2}\left[\frac{1}{7} + \frac{(0.0426)^2}{7}\right]$$

$$= 0.041,$$

$s_{R_B} = 0.204.$

Thus

$R_B = 0.426 \pm 0.204.$

The confidence interval for R_B is

$0.426 \pm z_{1-\alpha/2}(0.204).$

If $\alpha = 0.10$, then $R_L = 0.09$ and $R_U = 0.762.$

Remark. Notice that the sample sizes are fairly small for the confidence interval based on the normal approximation to be meaningful.

2.8. *Use of Covariates*

In the example of atrial natriuretic assay, we have considered sodium excretion due to injection of atrial extracts. It may be unreasonable to assume that the response is independent of body weight or weight of heart. If such a rule is adopted, one has to administer extreme doses to some human subjects, which are prohibited by the medical profession. For a wide range of body weights, proportional adjustment may be an oversimplification of the weight-tolerance relationship. Thus, a better procedure might be to do a regression (linear or nonlinear) analysis using body weight (or weight of heart, etc.) as the covariate. However, if the precision of estimates is not appreciably improved by using the regression analysis, then this method should be abandoned.

Appendix: Behrens-Fisher-Sukhatme Distribution

Let $X_1, X_2, \ldots, X_{m+1}$ be a random sample from normal (μ, σ^2), and $Y_1, Y_2, \ldots, Y_{n+1}$ be a random sample from normal (ν, τ^2) and assume that the X' and Y' values are mutually independent. We wish to test $H_0(\mu = \nu)$ against $H_1(\mu > \nu)$. Let $\bar{X}$ and $\bar{Y}$ denote the sample means and let $s = s_{\bar{X}}$ and $s' = s_{\bar{Y}}$. Consider

$$T = (\bar{X} - \mu)/s \text{ and } T' = (\bar{Y} - \nu)/s'.$$

Define

$$D = (\bar{Y} - \bar{X})/(s^2 + s'^2)^{1/2}.$$

If $\tan \theta = s/s'$, then under H_0,

$$D = (T's' + \nu - Ts - \mu)/(s^2 + s'^2)^{1/2}$$
$$= (T' - T\tan\theta)\cos\theta = T'\cos\theta - T\sin\theta, \text{ when } \mu = \nu.$$

Hence

$$P[D > d | H_0] = \iint f_m(t) f_n(t') dt \, dt'$$
$$ t'\cos\theta - t\sin\theta > d$$
$$= \int_{-\infty}^{\infty} \left[\int_v^{\infty} f_n(t') dt' \right] f_m(t) dt,$$

where $v = (d + t\sin\theta)/\cos\theta$, and $f_k(t)$ denotes the distribution of central t having k degrees of freedom. Carrying out numerical integrations, Sukhatme [1938] has tabulated the upper 5% points of D for m, n = 6, 8, 12, 24, ∞ and $\hat{\theta} = 0, 15°, 30°, 45°$. When $\theta = 0$, $D = T_n$, and when $\theta = 90°$, $D = T_m$, where T_k denotes a random variable having Student's t distribution with k d.f.

Questions

(1) Is the distribution of D symmetric about 0 when H_0 is true? (Easy to show that $P(D < -d|H_0) = P(D > d|H_0)$.)
(2) Can the characteristic function be found?
(3) Is it easy to evaluate the distribution of D when m = n?

3   Algebraic Dose-Response Relationships[1]

3.1.   *Indirect Assays*

Direct assays have their shortcomings, especially the bias introduced by the time-lag in producing the response. Even if this difficulty is overcome, it is not easy to administer the precise dosage that produces the desired response. For instance, it is almost impossible to determine individual tolerances of aphids for an insecticide.

In an indirect assay, specified doses are given each to a set of experimental units and the resulting responses are recorded. It is said to be a *quantal response* if the response recorded is 'all or nothing' like whether a death does or does not result. Other responses could be a change in the weight of a certain organ, the length of the subject's survival. In the latter case, the response is quantitative. Quantal responses are related to direct assays in this sense; if only the dosage that produced each subject's 'death' is recorded and all other responses are deleted, then quantal response becomes the response in direct assays. However, the quantitative assays are mathematically complicated. It will be more informative to establish a relationship between dose and the magnitude of the response produced by the dose. If this relationship is established for several preparations, equally effective doses may be estimated and by suitable statistical procedures the precision of the estimates of the relative potencies may be established.

3.2.   *Dose-Response Relationship in Terms of Regression*

Let z denote the dose and U denote the response it induces. Then if $u = E(U)$ we assume that there exists a functional relationship between u and z given by

$$u = h(z) \tag{3.1}$$

where $h(z)$ is a single-valued real function of z for all doses in the range of interest. If S and T are abbreviations for standard and test preparations, then the regression relationships for the two preparations can be denoted by

$$u_S = h_S(z) \quad \text{and} \quad u_T = h_T(z). \tag{3.2}$$

From (3.2) one can obtain the doses z_S and z_T that would induce the average response u. In other words, the doses z_S and z_T are equally effective. There may be some situations where z_S and z_T may not be determinable or may not be determined uniquely. However, in the case of analytical dilution assays, these

---

[1] Finney [1971b] served as a source for the material of this chapter.

uncertainties disappear since the two regression curves must be similar in shape. The relative potency z_S/z_T as a function of u can be plotted in the shape of a curve or an analytical expression can be obtained. In analytical dilution assays we make the following assumptions: (1) the response is produced solely by the factor x and nothing else like the impurity; (2) the response to be analysed is produced solely by the presence of factor x and not by any other substance, (3) the test preparation behaves like a dilution or a concentration of the standard preparation in a completely inert dilutent.

If the two preparations contain the same effective constituent (or the same effective constituents in fixed proportions) and all other constituents are without effect on u, so that one behaves like a dilution of the other, the potency ratio z_S/z_T must be independent of u, that is

$$z_S/z_T = \rho, \quad \text{hence} \tag{3.3}$$

$$h_T(z) = h_S(\rho z) \quad \text{for all } z, \tag{3.4}$$

where ρ is a constant, which is the condition for similarity that is satisfied by all dilution assays. In summary, the average response induced by a standard preparation has a dose-response regression function $u = h(z)$. A test preparation has the dose-response regression function $u = h(\rho z)$ where ρ denotes the relative potency, the number of units of effective constituent equivalent to 1 unit dose of the test preparation. An estimate R of ρ is sought from observations made on dose-response tests.

Remark. If the data arising from a series of tests cannot be described by the same form of regression function for both preparations, then either the assumption of similarity is incorrect or the conditions of testing have been different for the two preparations. It may happen that over a wide range of responses the two functions are nearly the same, although at extremes, they may differ significantly. Then the condition of similarity holds reasonably well. In a *comparative dilution assay*, its validity need not hold under conditions other than those of its estimation. However, in an *analytical dilution assay*, the estimate of potency should be independent of the experimental method and the assay technique employed.

3.3. *Preliminary Guess of the Regression Function*

We should have some idea of the form of h(z) before we estimate the unknown parameters occuring in h(z). A preliminary investigation involving the past data that has accumulated over the years will be needed in order to establish the form h(z), which may require the observed response at several values of z. Some people would separate the assay from this investigation involving the determination of h(z).

For most of the assays that we encounter in practice, we assume that h(z) is strictly monotone (either increasing or decreasing). However, we focus our attention on strictly increasing functions. That is, if $z_2 > z_1$ then $h(z_2) > h(z_1)$. It is quite possible that nonmonotonic h(z) can occur provided the condition of similarity holds. However, the dose of the standard preparation having the potency as some other dose of the test preparation is not unique. Thus, we may confine ourselves to the range of doses over which the function h(z) is monotone.

3.4. *Transformation Leading to Linear Relationships*

Let us assume that

$$u_S = h(\alpha + \beta z^\lambda) \tag{3.5}$$

where α, β and λ are unknown parameters and the form of h is completely specified and h is monotone. The limiting case of (3.5), when $\lambda = 0$, is given by

$$u_S = h(\alpha + \beta \log z). \tag{3.6}$$

This can be justified as follows. The rate of change in $\alpha + \beta z^\lambda$ is

$$\frac{d}{dz}(\alpha + \beta z^\lambda) = \lambda \beta z^{\lambda-1}.$$

The ratio of this rate of increase to its value at $z = 1$ is $z^{\lambda-1}$ which is increasing in λ for $z > 1$ and decreasing in λ for $z < 1$ and has a discontinuity at $\lambda = 0$. (To see this point more clearly consider $(\lambda - 1)\log z$ and study its behavior.) However, the rate of increase in $(\alpha + \beta \log z)$ is βz^{-1} and the ratio of this rate of increase to its value at $z = 1$ coincides with $\lim_{\lambda \to 0}(z^{\lambda-1})$.

In this sense (3.6) is a limiting form of (3.5) when $\lambda \to 0$. Equations (3.5) and (3.6) can easily be transformed into linear relationships. Let $x = z^\lambda$ or $x = \log z$, then

$$u_S = h(\alpha + \beta x). \quad \text{Now set} \tag{3.7}$$

$y_S = h^{-1}(u_S)$, yielding

$$y_S = \alpha + \beta x. \tag{3.8}$$

Then, the relation for the test preparation would be

$$\begin{aligned} Y_T &= \alpha + \beta \rho^\lambda x \quad (\lambda \neq 0) \\ Y_T &= \alpha + \beta \log(\rho + \beta x) \quad (\lambda = 0). \end{aligned} \tag{3.9}$$

For $\lambda \neq 0$, the lines intersect at $x = 0$ and R, the estimate of ρ is the $(1/\lambda)$-th power of the ratio of the slopes of the two linear equations. If $\lambda = 0$, the lines are parallel and R is equal to the antilog of the difference of the intercepts of the two linear equations divided by the slope. In the literature, $x = z^\lambda$ is called the dose-metameter and $y = h^{-1}(u)$ is called the response metameter. Thus, if λ is known, since h is a known function, the usual method of estimating potency from data on two preparations is to express doses and responses metametrically and to fit to two straight lines subject to the restriction that the lines either intersect at x = 0 or are parallel. Thus, the value of ρ will be a function of λ. Hence, the preliminary investigation consists of determining the form of h and the value of λ. The most common values of λ are 0 and 1, other simple values of λ are ½, ⅓, −1, −½. Potential candidates for $h^{-1}(u)$ are u, log u, 1/u or $u^{1/2}$. From tests with a large dosis range, the relation between u and x would be determined empirically by plotting $\bar{u}$, the mean response against x and fitting a regression relationship by eye or by some other procedure. If h(z) = z and λ = 0, then the metametric regressions given by

$$Y_S = x$$
$$Y_T = \log(\rho + x) \tag{3.10}$$

yields the standard regression in which case α = 0 and β = 1.

3.5. *Nonlinear Regression*

Emmens [1940] proposes that a logistic function might provide a good fit to the regression of a quantitative response on dose when the range of responses is too great for a simple linear regression of u on x = log z to hold. We can write this function as

$$u = (½)H[1 + \tanh(\alpha + \beta x)], \quad \text{where} \tag{3.11}$$

$\tanh \theta = [\exp(2\theta) - 1]/[\exp(2\theta) + 1]$.

Emmens [1940] finds (3.11) to be a good representation for a number of data. If the constant H is known or could reasonably be estimated from preliminary investigations, the metametric transformation

$$y = \tanh^{-1}\left(\frac{2u - H}{H}\right) \tag{3.12}$$

reduces equation (3.11) to the form of equation (3.8). Typically H will be unknown and hence must be estimated, for instance by the method of maximum likelihood. Also one could use

$$u = H \int_{-\infty}^{\alpha + \beta x} (2\pi)^{-1/2} \exp(-t^2/2) dt = H\Phi(\alpha + \beta x) \tag{3.13}$$

where the logistic has been replaced by the normal. Neither of the models seem to embrace all the situations because in either of the models the regression curve flattens and goes to zero as x → −∞ and goes to H as x → ∞. The flat parts of the curve will not be appropriate for the assay problems (because small changes in response correspond to large dose differences). Furthermore, the variance of the response at the extremes of the dose may depend on the response itself. Sometimes an adjustment to the regression equation will improve matters. For instance, we can take

$$u_S = \alpha + \beta x + \gamma x^2, \quad \text{which yields} \tag{3.14}$$

$$u_T = \begin{cases} \alpha + \beta(\log\rho + x) + \gamma(\log\rho + x)^2, & \lambda = 0 \\ \alpha + \beta\rho^\lambda x + \gamma\rho^{2\lambda} x^2, & \lambda \neq 0. \end{cases} \tag{3.15}$$

However, the estimation of the parameters α, β, γ, λ may be troublesome. Consider the following example.

Example

Table 3.1 pertains to percentage survival Y of the beetle *Tribolium castaneum* at four densities (x denotes the number of eggs per gram of flour medium).

Table 3.1. Comparison of fitness characters and their responses to density in stock and selected cultures of wild type and black *Tribolium castaneum* [Sokal, 1967]

	Density			
	$x_1 = 5/g$	$x_2 = 20/g$	$x_3 = 50/g$	$x_4 = 100/g$
Survival in proportions	0.775	0.862	0.730	0.640
	0.725	0.844	0.725	0.585
	0.875	0.800	0.725	0.584
	0.775	0.763		
	0.875			
$X_i = \log_{10} x_i$	0.699	1.301	1.699	2.000
n_i	5	4	3	3
$\sum_j Y_{ij}$	4.025	3.269	2.18	1.809
$\bar{Y}$	0.805	0.817	0.727	0.603

$\bar{X} = (0.699 + 1.301 + 1.699 + 2.000)/4 = 5.699/4 = 1.425$
$\bar{Y} = (4.025 + 3.269 + 2.18 + 1.809)/15 = 11.283/15 = 0.752$
The fitted regression equation is $Y = 0.929 - 0.134X$.
s_a = standard error of the intercept = 0.049;
s_b = standard error of the slope = 0.035.

Table 3.2. Anova table for *Tribolium castaneum* data

Source	d.f.	Sum of squares	Mean square	F value	Prob > F
Linear regression	1	0.0673	0.0673	28.45	0.0002
Deviation from linear regression	2	0.0323	0.0162	6.82	0.012
Error	11	0.0260	0.0024		
Total	14	0.1256			

From table 3.2 we infer that the linear fit is significant and the deviation from the linear fit is not significant at 1% level of significance.

3.6. *Heterogeneity of Variance*

If the variance of the response U for specified z is dependent upon $u = EU$, a transformation is necessary before the analysis is performed.
If

$$\sigma_U^2 = \psi(u), \quad EU = u$$

then the proposed transformation is

$$U^* = \int^U \psi^{-1/2}(v)dv.$$

Then

$$dU^* = dU/\psi^{1/2}(U).$$

Hence as a first approximation

$$\sigma_{U^*}^2 = \sigma_U^2/\psi(u) = 1.$$

Bartlett [1937] considers such transformations. The transformed response U^* will have homoscedasticity of variances. In the following we shall consider some special cases of ψ.

(1) $\sigma_U^2 = u$ implies that $U^* = 2U^{1/2}$

(2) $\sigma_U^2 = u^2$ implies that $U^* = \ln U$

(3) $\sigma_U^2 = u(H - u)/Ha$ implies that $U^* = \sin^{-1}[(u/H)^{1/2}]$, $a^{-1/2} = 2H^{1/2}$.

Table 3.3. Test for homoscedasticity on the *Tribolium castaneum* data

Dose level	$\bar{Y}_i$	d.f. = f_i	Sample variance s_i^2	$\ln s_i^2$
0.699	0.805	4	0.0045	−5.4037
1.301	0.817	3	0.0020	−6.2215
1.699	0.727	2	$0.83(10^{-5})$	−11.6952
2.000	0.603	2	0.001027	−6.8811

$\sum f_i = 4 + 3 + 2 + 2 = 11$
$s^2 = \sum f_i s_i^2 / \sum f_i = 0.0260145/11 = 0.002365$
$\ln s^2 = -6.0470$
$\sum f_i \ln s_i^2 = -77.4320.$

Using the test of Bartlett [1937] given by

$$\chi^2_{k-1} = [f \ln s^2 - \sum f_i \ln s_i^2]/c, \quad \text{where}$$

$$c = 1 + [3(k-1)]^{-1}\left[\sum_1^k f_i^{-1} - f^{-1}\right]$$

$$f = \sum_1^k f_i \quad (k = 4),$$

we obtain

$$c = 1 + \frac{1}{9}\left[\frac{1}{4} + \frac{1}{3} + \frac{1}{2} + \frac{1}{2} - \frac{1}{11}\right] = 1 + \frac{1.4924}{9} = 1.1658.$$

Hence

$$\chi^2_3 = [77.4320 + 11(-6.0470)]/1.1658$$

$$= 10.9152/1.1658 = 9.36, \quad \text{and}$$

$$P(\chi^2_3 > 9.36) = 0.025.$$

So, at 1% level of significance we accept the null hypothesis of homogeneity of variance.

Since the response is bounded, the angular transformation given by case (3) may be appropriate. So, using the transformation $Z = \arcsin\sqrt{Y}$, we obtain the following data:

Table 3.4. Survival data in degrees

Density			
$x_1 = 5/g$	$x_2 = 20/g$	$x_3 = 50/g$	$x_4 = 100/g$
61.68	68.21	59.69	53.13
58.37	66.72	58.37	49.89
69.30	63.44	58.37	49.82
61.68	60.84		
69.30			

We also obtain the following table:

Dose	$\bar{Z}$	d.f.	s_i^2	$\ln s_i^2$
0.699	64.07	4	0.007498	−4.8930
1.301	64.80	3	0.003309	−5.7112
1.699	58.81	2	$0.105(10^{-4})$	−11.4633
2.000	50.95	2	0.001083	−6.8282

Further computations yield:

$s^2 = \sum f_i s_i^2 / \sum f_i = 0.003828$

$\sum f_i \ln s_i^2 = -73.2887$

$\ln s^2 = -5.5654$

$\chi_3^2 = [11(-5.5654) + 73.2887]/c$

$\quad = 12.0688/c = 10.35$

$P[\chi_3^2 > 10.35] = 0.015.$

So, at 1% level of significance, we still accept the hypothesis of homogeneity of variances. The arc sin transformation stabilizes variances if there are large numbers of values close to 0 and 1.

Next let us fit the linear regression on the transformed data. We obtain the estimated regression line as

$$Z = 1.2602 - 0.1545X \qquad (3.16)$$

s_a = standard error of the intercept = 0.0578

s_b = standard error of the slope = 0.0409.

The following Anova table results:

Table 3.5. Anova table for the transformed data

Source	d.f.	Sum of squares	Mean square	F value	Prob > F
Linear regression	1	0.0894	0.0894	23.36	0.0005
Deviation from linear regression	2	0.0396	0.0198	5.17	0.026
Error	11	0.0421	0.0038		
Total	14	0.1711			

From table 3.5 we infer that the linear fit is significant and the deviation from the linear fit is not significant at 1% level of significance.

3.7. *Maximum Likelihood Estimates of Parameters*

There are several methods of estimating the parameters available in the literature, namely (1) the method of moments, (2) the method of maximum likelihood, (3) the method of minimum χ^2 and (4) the method of generating best asymptotically normal (BAN) estimates.

Among these, we prefer the method of maximum likelihood because (a) maximum likelihood estimators (mle) are asymptotically unbiased, (b) they are consistent, (c) they are asymptotically normal, (d) they are asymptotically efficient, (e) they are functions of the sufficient statistic, (f) they are invariant in the sense that if g is a continuous function of θ, the unknown parameter, then mle of $g(\theta)$ is equal to g evaluated at the mle of θ, and (g) mle can be interpreted as the mode of the posterior density of θ when (irregular) uniform prior density is assumed for θ.

3.8. *Maximum Likelihood: Iterative Scheme*

Since we will be dealing with arithmetic means of responses, it is not unreasonable to assume the normal distribution for the response. Let U be $\stackrel{d}{=}$ normal (u, σ^2). Then the log likelihood function is[2]

$$L = \text{constant} - \sum (U_i - u_i)^2 / 2\sigma^2 \tag{3.17}$$

where the constant involves σ and not other parameters and the summation is over all the responses. Notice that u_i involves the unknown parameters α and β [in fact, $u_i = h(\alpha + \beta x_i), i = 1, 2, \dots$]. Let a and b denote the mle values of α and β, respectively, and be the solutions of the likelihood equations:

$$\partial L / \partial \alpha = 0, \quad \partial L / \partial \beta = 0. \tag{3.18}$$

Let a_1, b_1 be some simple approximations to a and b obtained by some method (such as the eye test etc.). Then let $a_2 = a_1 + \delta a_1$, $b_2 = b_1 + \delta b_1$ where δa_1 and δb_1 are given by

$$\begin{aligned}
\frac{\partial L}{\partial \alpha_1} + \delta a_1 \frac{\partial^2 L}{\partial \alpha_1^2} + \delta b_1 \frac{\partial^2 L}{\partial \alpha_1 \partial \beta_1} &= 0 \\
\frac{\partial L}{\partial \beta_1} + \delta a_1 \frac{\partial^2 L}{\partial \alpha_1 \partial \beta_1} + \delta b_1 \frac{\partial^2 L}{\partial \beta_1^2} &= 0.
\end{aligned} \tag{3.19}$$

Suffix 1 on α and β indicates that the partial derivatives are evaluated at a_1 and b_1. Consider

$$\partial L / \partial \alpha = \sigma^{-2} \sum_i (U_i - u_i) \partial u_i / \partial \alpha. \tag{3.20}$$

In iterative solutions of mle, one can replace the second partial derivatives of L by their expected values since the expected values can be tabulated and by the strong laws of large numbers, the 2nd derivatives converge to their expected values.

---

[2] We assume that all the responses are arranged in terms of the elements of a single vector. That is, not all x values are distinct and there may be several U values at the same dose level x_i.

By substituting $U_i = u_i$ after differentiating (3.20), we obtain

$$E(\partial^2 L/\partial \alpha^2) = -\sigma^{-2}\sum(\partial u_i/\partial \alpha)^2$$
$$E(\partial^2 L/\partial \alpha \partial \beta) = -\sigma^{-2}\sum(\partial u_i/\partial \alpha)(\partial u_i/\partial \beta) \quad \text{and} \tag{3.21}$$
$$E(\partial^2 L/\partial \beta^2) = -\sigma^{-2}\sum(\partial u_i/\partial \beta)^2.$$

Write

$$\partial u/\partial y = h'(y) \quad \text{where } y = \alpha + \beta x \tag{3.22}$$

and define the weight function W by

$$W = [h'(y)]^2. \tag{3.23}$$

Then, one can rewrite (3.21) as

$$E(\partial^2 L/\partial \alpha^2) = -\sigma^{-2}\sum W_i(\partial y/\partial \alpha)^2 = -\sigma^{-2}\sum W_i$$
$$E(\partial^2 L/\partial \alpha \partial \beta) = -\sigma^{-2}\sum W_i x_i \quad \text{and} \tag{3.24}$$
$$E(\partial^2 L/\partial \beta^2) = -\sigma^{-2}\sum W_i x_i^2.$$

Corresponding to the initial estimates a_1 and b_1 the expected value of the response metameter Y is given by y_1 where

$$y_1 = a_1 + b_1 x. \tag{3.25}$$

Using (3.23)–(3.25) in (3.19) we have

$$\delta a_1 \sum W_{1i} + \delta b_1 \sum W_{1i} x_i = \sum_i W_{1i}(U_i - u_i)/h'(y_{1i}),$$
$$\delta a_1 \sum W_{1i} x_i + \delta b_1 \sum W_{1i} x_i^2 = \sum_i [W_{1i} x_i (U_i - u_i)/h'(y_{1i})]. \tag{3.26}$$

Now add $\sum W_{1i} y_{1i} = \sum W_{1i}(a_1 + b_1 x_i)$ to both sides of the first equation and $\sum W_{1i} x_i y_{1i}$ to both sides of the second equation in (3.26). If the working response Y is defined by

$$Y = y + (U - u)[h'(y)]^{-1} \tag{3.27}$$

Equation (3.26) can be rewritten as

$$a_2 \sum W_{1i} + b_2 \sum W_{1i} x_i = \sum W_{1i} Y_{1i}$$
$$a_2 \sum W_{1i} x_i + b_2 \sum W_{1i} x_i^2 = \sum W_{1i} x_i Y_{1i}. \tag{3.28}$$

Now (3.28) implies that

$$y_2 = a_2 + b_2 x \tag{3.29}$$

can be obtained as the weighted linear regression equation of Y_1 on x. Now

$$\bar{x}_1 = \sum W_{1i} x_i / \sum W_i, \tag{3.30}$$
$$\bar{Y}_1 = \sum W_{1i} Y_{1i} / \sum W_{1i}, \quad \text{and}$$

$$\sum_{xx} = [\sum W_i x_i^2 - (\sum W_i x_i)^2]/\sum W_i,$$

$$\sum_{xY} = [\sum W_i x_i Y_i - (\sum W_i x_i)(\sum W_i Y_i)]/\sum W_i.$$

Then

$$b_2 = \sum_{x_1, Y_1} \bigg/ \sum_{x_1 x_1}$$

$$a_2 = \bar{Y}_1 - b_2 \bar{x}_1. \tag{3.31}$$

Now the process can be iterated with a_2, b_2 replacing a_1 and b_1. The iteration can be stopped when one approximation does not differ appreciably from the immediately preceding one.

Remark. Notice that the working response Y given by (3.27) does not coincide with $h^{-1}(U)$, unless $h(u) = u$. However, we always have $EY = y = h^{-1}(u)$, $u = EU$. U represents the original response, or the value resulting after a scedasticity transformation. Unless otherwise stated, Y denotes the working response defined by (3.27).

Finney [1949b, 1952] extended the method of iterated maximum likelihood estimation to quantal responses.

Towards the variances, we have

$$\text{var}(\bar{Y}) = \sigma^2/\sum W_i$$

$$\text{var } b = \sigma^2 \bigg/ \sum_{xx} \tag{3.32}$$

where $W_i = \lim_j W_{j,i}$ ($i = 1, \ldots$) and j denote the stage of iteration (that is, they are the limiting values.) Also, W_i and $\sum_{xx}$ will be close to the values obtained at the last iteration. The unknown σ^2 will be replaced by the estimate s^2 based on the residual sum of squares, namely, the sum of deviations of the individual values of U about the estimated regression equation.

In summary, it is desirable to have a table of values of the metametric transformation that is used, i.e. a table of values of $y = h^{-1}(u)$ is required. By a table of the weighted coefficients given by equation (3.23), the minimum working response

$$y_0 = y - u[h'(y)]^{-1}, \tag{3.33}$$

and the range

$$A = 1/h'(y) \tag{3.34}$$

as functions of y can be made. Notice that the suffixes indicating the stage of iteration are dropped in equations (3.33) and (3.34). First we read the empirical response corresponding to each observed U from the table and then plot these against x. A straight line is drawn by eye through these points. Some allowance for unequal weights can be made when positioning the line. Expected responses y are read from the lines which correspond with the observed x values. From the second table, the weighting coefficient for each y is read and the corresponding working response is formed as

$$Y = y_0 + UA. \tag{3.35}$$

Then the weighted regression of Y on x is calculated. Now with this new set of values for expected responses, a second iteration can be performed.

This process of iteration is stopped when the pair of coefficients a_{r-1}, b_{r-1} differs very little from the pair a_r, b_r and the last set is regarded as the mle values of α and β. The process of iteration converges rapidly and it should not take more than two or three iterations, provided that the initial values, a_1, b_1 are chosen judiciously.

For quantitative responses, if the original observations are homoscedastic or if a scedasticity transformation has been performed, then

$$y = u, \quad h'(y) = 1 \quad \text{and} \quad W = 1.$$

Hence

$$Y = U$$

and the iterative process described earlier simplifies to calculating the unweighted linear regression and clearly one cycle suffices. In several practical situations, one can reasonably assume homogeneity of variances and the linearity of the regression.

3.9. *Estimation of the Relationship for the Standard Preparation*

Suppose that in an assay of a cod liver oil for its vitamin D content, 10 rats received a dose of 50 mg and the average degree of curing initial rickets was 2.10 or a score of 8.40 on the scale of quarter units. The equivalent log dose having an expected response of 8.40 is given by

$$x = (8.40 - 5.89)/17.14 = 0.146.$$

Thus, the 50 mg cod liver oil has the same effect as antilog $0.146 = 1.40$ IU of vitamin D. In other words, 1 g of cod liver oil is estimated to contain 28.0 IU vitamin D.

Remark. Due to random fluctuations in experimental conditions within a laboratory, it is not good to put too much faith in the estimate of the potency. Thus, in general, a response once determined cannot indefinitely be used. It should be emphasized that the response curves for the standard as well as the test preparation should be based on data gathered in the same laboratory under identical experimental conditions.

3.10. Estimation of the Slope

Although the response regression might shift from day to day and under varied experimental conditions, it is not unreasonable to assume that the slope, namely the increase in response to per unit increase in log dose, may remain fairly constant. Also the foregoing method of estimating the standard slope should be restricted to linear regression relation between log dose and response.

For instance, let a group of 8 rats receive 5 IU vitamin D each and show a mean response of 12.25. Also use the data on 10 rats that received 50 mg of cod liver oil and yielded an average response of 8.40 (table 3.6). If x_S, x_T denote log doses of S and T, respectively, and $\bar{Y}_S$ and $\bar{Y}_T$ the corresponding mean responses, the logarithm of the estimate of relative potency is

$$M = x_S - x_T - (\bar{Y}_S - \bar{Y}_T)/b = (a_S - a_T)/b, \text{ where}$$

$\bar{Y}_S = a_S + bx_S$ and

$\bar{Y}_T = a_T + bx_T$

Hence

R = antilog M.

For the above data

$M = \log 5 - \log 50 - (3.85)/77.14 = -1 - 0.223 = \bar{2}.775,$

yielding R = 0.0596. That is, 1 g of the cod liver oil is estimated to contain 59.6 IU vitamin D.

Table 3.6. Hypothetical assay of vitamin D

	Response to		Equivalent angles	
	5 IU S	50 mg T	5 IU S	50 mg T
	15	4	52	24
	10	10	40	40
	18	12	60	45
	6	7	30	33
	9	5	38	27
	14	5	50	27
	12	9	45	38
	14	14	50	50
		10		40
		8		35
n	8	10	8	10
$\bar{y}$	12.25	8.40	45.62	35.90
$\sum y$	98	84	365	359

Reproduced from Finney [1971b, p. 93].

3.11. *Estimation Based on Simultaneous Tests*

In order to overcome the objections to the standard-curve and standard-slope methods of estimating potency one can conduct simultaneous experiments on both the preparations under identical conditions, using two or more doses of each preparation. The required quantities can be evaluated from the present assay data. The present assay, because of a fewer number of observations made, leads to estimates of regression coefficients that are less precise; however, they will be current and up-to-date. Empirical responses for both the standard and test preparations should be plotted against dose levels x and two preliminary regression lines be drawn subject to the constraint that either they intersect at x = 0 or they are parallel. The weighted regression calculations are carried out in order to improve the approximations to the estimates of all parameters, maintaining the constraint. Finally, ρ is estimated either from the ratio of the slopes or from the difference of the intercepts of the parallel lines on the response axis. If simultaneous trials are carried out on 3 or more doses of each preparation, then a test for deviations from linearity can be constructed.

Appendix: Regression Models [Lindgren, 1976]

Linear Regression Model

Let

$$Y_{ij} = \gamma_i + \varepsilon_{ij} \quad \text{or}$$

$$Y_{ij} = \alpha + \beta(x_i - \bar{x}) + \varepsilon_{ij} \quad (j = 1, \ldots, n_i, i = 1, \ldots, k)$$

where not all x values are the same. We assume that

$$E\varepsilon_{ij} = 0 \quad \text{and} \quad \text{var}\,\varepsilon_{ij} = \sigma^2.$$

Hence we have

$$\hat{\alpha} = \bar{Y}, \hat{\beta} = \frac{\sum(x_i - \bar{x})(\bar{Y}_i - \bar{Y})}{\sum(x_i - \bar{x})^2}, \quad \text{where}$$

$$\bar{Y}_i = \sum_{1}^{n_i} Y_{ij}/n_i \;(i = 1, \ldots, k), \quad \text{let } \tilde{Y}_i = \hat{\alpha} + \hat{\beta}(x_i - \bar{x}), \quad \text{and } n = n_1 + \cdots + n_k.$$

Let $\hat{\sigma}^2 = n^{-1}\sum\sum(Y_{ij} - \tilde{Y}_i)^2$.

We wish to test

$H_0: \gamma_i = \alpha + \beta(x_i - \bar{x})$ versus H_A: the (x_i, γ_i) are (not collinear).

Consider

$$\sum\sum(Y_{ij} - \tilde{Y}_i)^2 = \sum\sum(Y_{ij} - \bar{Y}_i)^2 + \sum_{i=1}^{k} n_i(\bar{Y}_i - \tilde{Y}_i)^2,$$

that is, residual sum of squares = pure error sum of squares + lack of fit sum of squares. If we define F by

$$F = \frac{\sum n_i (\bar{Y}_i - \tilde{Y}_i)^2 / (k-2)}{\sum\sum (Y_{ij} - \bar{Y}_i)^2 / (n-k)}$$

then F, under H_0, is distributed as Snedecor's F with degrees of freedom $k - 2$ and $n - k$, when the ε_{ij} are independent and are normally distributed.

Step-Wise Regression (Forward Selection Procedure)

Let the model be $Y_{ij} = h(x_i) + \varepsilon_{ij}$ ($j = 1, \ldots, n_i$, $i = 1, \ldots, k$). Assume that the ε_{ij} are independent and normal $(0, \sigma^2)$. We wish to test

H_1: $h(x) = \beta_1$

H_2: $h(x) = \beta_1 + \beta_2 x$

$\vdots$

H_k: $h(x) = \sum_{i=1}^{k} \beta_i x^{i-1}$.

In general, if model 1 involves m_1 parameters and model 2 involves m_2 parameters, with $m_1 < m_2$, then the likelihood ratio for model 1 versus model 2 is

$$\Lambda = \sup L(\beta_1 \ldots \beta_{m_1}; \sigma^2) / \sup L(\beta_1 \ldots \beta_{m_2}; \sigma^2)$$
$$= (\hat{\sigma}_1^2 / \hat{\sigma}_2^2)^{-n/2}$$

where L denotes the likelihood of the parameters,

$$n\hat{\sigma}_1^2 = \sum_i \sum_j [Y_{ij} - \hat{h}_1(x_i)]^2 = Q_1,$$

$$n\hat{\sigma}_2^2 = \sum\sum [Y_{ij} - \hat{h}_2(x_i)]^2 = Q_2, \quad n = \sum_1^k n_i,$$

and $\hat{h}_1$ and $\hat{h}_2$ are the least squares estimates of the regression function under models 1 and 2 respectively. According to the partition theorem (Cochran's theorem, Lindgren [1976, p. 525]), Q_1/σ^2 and Q_2/σ^2 are χ^2 with $n - m_1$ and $n - m_2$ d.f. Also $Q_1 = (Q_1 - Q_2) + Q_2$.

We have the following facts from the general theory of linear models:

(1) $(Q_1 - Q_2)/\sigma^2$ is distributed as $\chi^2_{m_2 - m_1}$ under model 1,
(2) Q_2/σ^2 is distributed as $\chi^2_{n - m_2}$ under model 2,
(3) Q_2 and $Q_1 - Q_2$ are independent under model 2.

Then

$$T = \frac{(Q_1 - Q_2)/(m_2 - m_1)}{Q_2/(n - m_2)} \stackrel{d}{=} F_{m_2 - m_1, n - m_2}$$

when the null hypothesis, namely model 1, is true. Notice that the alternative hypothesis is that the true model is included in model 2, which is more complicated than model 1.

In particular, if we wish to analyze the hierarchy of hypotheses given earlier, where $H_1 \subset H_2 \subset \cdots \subset H_k$, and $H \subset K$ denotes that model H can be deduced as a special case of model K. Then, we have the following Anova table:

Source	Sum of squares	d.f.
Fitting H_2 after H_1	$Q_1 - Q_2$	1
$\vdots$	$\vdots$	$\vdots$
Fitting H_k after H_{k-1}	$Q_{k-1} - Q_k$	1
Error	Q_k	$n - k$
Total (fitting H_1)	Q_1	$n - 1$

Remark. If the predictor variables are more than 1, and one is interested in fitting a multiple linear regression, there are two ways of going about it: (1) backward elimination (BE) procedure and (2) forward selection (FS) procedure. Then the regressor variables need to be ordered in terms of their importance; for instance, with respect to their partial correlations with the predicted variable [Draper and Smith, 1981].

4   The Logit Approach[1]

4.1.   *Introduction*

An action curve in pharmacology describes the amount of the response to any physical or chemical stimulation expressed as a percentage of the maximum obtainable in that particular biological system. Also, the action curve is invariably a sigmoid, in the sense that the plot of the percentage of dead organisms against some function of dosage is S-shaped. The change in percentage kill per unit of the abscissa is smallest near mortalities 0 and 100% and largest near 50%. Thus, the dose mortality curve describes the variation in susceptibility among individuals of a population. One can reasonably expect this susceptibility to follow the cumulative normal curve. Then the question is what function of the dosage should be taken as the absissa. Typically, the dosages increase and decrease in equal additive increments. Galton [1879] points out that the variation between individuals in their susceptibility to biological material exhibits a geometrical rather than an arithmetical distribution and this has been confirmed by several investigators of toxic substances. Thus, the logarithm of the dosage can be viewed as an index to the inherent susceptibility of the individual to the drug or poison. The Weber-Fechner law [Clark, 1933] implies that the concentration of the poison in the dose is proportional to the amount of poison fixed by the tissues of the experimental animal; although there is no evidence to support such a relationship. Since the susceptibility of an animal can be viewed as the average susceptibility of its component cells, it is probable that the average susceptibility of an animal is normally distributed. Thus, Galton [1879] and others surmised that the tolerance curve of an experimental unit is approximately normal; that is, plotting the proportion of positive responses on normal probability paper against log-dose would result in a straight line. Several methods, both graphical and analytical, are available for fitting this line from which the EDp can be estimated for $0 < p < 1$.

The fixation of a drug or poison seems to be a phenomena of adsorption and two basic formulae describing this process were proposed by Freundlich [1922] and Langmuir [1917]. (See, for instance, Bliss [1935, pp. 142–143] or Clark [1933, p. 4].) If x denotes the concentration of a drug, y denotes the amount fixed in this organism, m is the mass of adsorbing constituents within the organism, and k and n are constants, then Freundlich's empirical formula is given by $kx^{1/n} = y/m$.

---

[1] Ashton [1972] served as a source for part of this chapter.

Since m will be constant from animal to animal, we have $\log x = n \log y + K'$ (where, typically $n = 0.5$) which establishes a linear relation between the log-dosage and the logarithm of the amount fixed by the cells of the animal.

In several instances, Langmuir's [1917] adsorption formula fits more satisfactorily the biological data on the fixation of drugs than the Freundlich's formula. Langmuir's hyperbolic adsorption formula is given by $kx^n = y/(100 - y)$ [Suits and Way, 1961, pp. 95, 445], where x denotes the concentration of the drug, y the percentage of the maximum amount of drug which can be fixed by the cell, n is determined by the molecular state of the fixed drug as compared with its state before adsorption and is usually 1 or 2; and k is constant. However, y cannot directly be measured. Since the changes observed should be a direct result of the fixation of the drug by the cells, y is estimated by the response, namely, the percentage kill of animals at the dose level. Although Langmuir's formula may cause some problems at the lower dose levels, in many cases, the log dose-mortality line agrees satisfactorily with higher kills. At lower dose levels, the straight line need to be bent up if it is to fit to the entire range of observations. For example, Shepard [1934] obtained a satisfactory linear fit for dose versus the 'logit', namely logarithm of the ratio of percent killed to percent surviving by means of the equation

$kx^n = y/(100 - y)$ with $\log k = -18.2$ and $n = 10.2$.

As pointed out by Clark [1933, p. 5], 'it must be remembered that a formula is merely a convenient form of a shorthand and is an aid to and not a substitute for reason'.

Thus, the logit method has a scientific justification. Other nice features of the logit method are easy calculation of the parameter estimates and the theoretical appeal due to the existence of sufficient statistics for the unknown parameters. Hence, the logit method is emphasized in most of the recent literature on estimating EDp.

In this chapter, we dwell upon the other approach, based on the logistic distribution. Let

$x = \log_{10}(\text{dose}) = \log_{10}(z)$.

Then the dose response relation after a metametric transformation of the response is

$y = \alpha + \beta x$

and the estimated relation is $y = a + bx$ where a and b are the usual regression estimates. If Y_S and Y_T denote the mean responses of the two groups of experimental units for the standard and test preparations, then

$x_S = (Y_S - a)/b$, $x_T = (Y_T - a)/b$.

The Logit Approach

So

$$x_T - x_S = (Y_T - Y_S)/b$$

gives (on the log scale) the excess potency of the test preparation over that of the standard and antilog $(x_T - x_S)$ gives the ratio of the number of standard units in the doses of the two preparations. Let X_S and X_T, respectively, denote the log doses of the standard and test preparations measured in milligrams. Then, let

$M = \log_{10}$ (potency of test/potency of standard).

The expression for M can be obtained in the following way. The equations of the parallel dose-response curves are given by

$$Y - Y_S = b(X - X_S)$$
$$Y - Y_T = b(X - X_T).$$

The horizontal difference between the lines yields the difference in $\log_{10}$ (dose) for equal responses, and thus $\log_{10}$ (potency ratio). When $Y = 0$

$$X = X_S - Y_S b^{-1} \quad \text{and} \quad X = X_T - Y_T b^{-1}, \quad \text{yielding}$$

$$M = X_S - X_T + (Y_T - Y_S) b^{-1}.$$

After using the formula

$$\operatorname{Var}(U/V) \doteq [(EV)^2 \operatorname{var} U + (EU)^2 \operatorname{var} V - 2(EU)(EV)\operatorname{Cov}(U,V)]/(EV)^4,$$

we obtain

$$S_M^2 \doteq \frac{s^2}{b^2}(n_S^{-1} + n_T^{-1}) + (Y_T - Y_S)^2 s_b^2 b^{-4}, \quad \text{where}$$

$$s_b^2 = s^2 / [\sum n_{Si}(x_{Si} - \bar{x}_S)^2 + \sum n_{Ti}(x_{Ti} - \bar{x}_T)^2]$$

and n_S and n_T denote the number of experimental units in the two groups. The second term goes to zero when either b is taken to be exact or when EY_S and EY_T are equal. As pointed out, using Fieller's theorem, one can set up confidence intervals for the potency.

4.2. *Case when the Dose-Response Curve for the Standard Preparation Is Unknown*

At least two dose levels of the two preparations are required for calculating the respective slopes. If the slopes are not significantly different, the pooled value of b will be used for both the lines. Let k dose levels be used and let

$X_S = \log_{10}$ (dose) for the standard preparation,

$X_T = \log_{10}$ (dose) for the test preparation,

$Y_{S,i}$ = mean response of $n_{S,i}$ units receiving the i-th dosage of the standard preparation, and

$Y_{T,i}$ = mean response of $n_{T,i}$ units receiving the i-th dose of the test preparation.

Let the mean values be denoted by

$$\bar{X}_S = \sum_1^k n_{S,i} X_{S,i} \Big/ \sum n_{S,i}, \quad \bar{Y}_S = \sum n_{S,i} Y_{S,i} / \sum n_{S,i}.$$

and analogous definitions of $\bar{X}_T$ and $\bar{Y}_T$ hold, where the summations are taken over the k levels. Then the equation of the two lines are

$$Y - \bar{Y}_S = b(X_S - \bar{X}_S)$$
$$Y - \bar{Y}_T = b(X_T - \bar{X}_T), \quad \text{where}$$

$$b = \frac{\sum n_{S,i} Y_{S,i}(X_{S,i} - \bar{X}_S) + \sum n_{T,i} Y_{T,i}(X_{T,i} - \bar{X}_T)}{\sum n_{S,i}(X_{S,i} - \bar{X}_S)^2 + \sum n_{T,i}(X_{T,i} - \bar{X}_T)^2}.$$

Then the potency ratio is given by

$$M = \bar{X}_S - \bar{X}_T + (\bar{Y}_T - \bar{Y}_S) b^{-1} \quad \text{and}$$

$$s_b^2 = s^2 \left[\sum n_{S,i}(X_{S,i} - \bar{X}_S)^2 + \sum n_{T,i}(X_{T,i} - \bar{X}_T)^2 \right]^{-1},$$

$$s_M^2 = s^2 b^{-2}(n_S^{-1} + n_T^{-1}) + (\bar{Y}_T - \bar{Y}_S)^2 s_b^2 b^{-4}$$

$$n_S = \sum n_{S,i}, \quad n_T = \sum n_{T,i}, \quad \text{where}$$

s^2 = (sum of squared deviations from group means)/$[\sum(n_{S,i} + n_{T,i} - 2)]$.

Note that as s_b/b increases, the contribution made by the second term to s_M increases.

Example

Consider the following (artificial) data pertaining to weight gain in pounds by piglets in a fixed duration of time when the standard and test preparation of a certain diet are administered to them.

	Gain in weight, y			
	standard preparation		test preparation	
	1 unit	2 units	1 unit	2 units
	3	4	3	4
	2	3	3	4
	4	5	2	6
	2	4	4	5
	3	5	4	4
n_i	5	5	5	5
$\sum Y_i$	14	21	16	23
$\bar{Y}_i$	•2.8	4.2	3.2	4.6

Computations yield:

Let $\log_{10} x = X$

$\bar{X}_S = 0.1505$, $\bar{X}_T = 0.1505$

$\bar{Y}_S = (2.8 + 4.2)/2 = 3.5$

$\bar{Y}_T = (3.2 + 4.6)/2 = 3.9$

$a_S = 2.8$, $a_T = 3.2$, $s_{a_S} = 0.32$, $s_{a_T} = 0.32$

$b = 4.65$, $s_b = 1.23$

$M = (0.1505 - 0.1505) + (3.9 - 3.5)/4.65 = 0.086$

$R = $ potency ratio $= 10^{0.086} = 1.219$

$s^2 = 0.682$, $s = 0.826$

$$s_M^2 = (4.65)^{-2}\left[\frac{0.6823 \times 2}{10} + \left(\frac{0.4 \times 1.23}{4.65}\right)^2\right]$$

$$= (4.65)^{-2}[0.136 + (.1058)^2]$$

$$= (4.65)^{-2}(0.1472)$$

$s_M = 0.082.$

4.3. *Quantal Responses*

If the percentage of subjects responding is plotted against $\log_{10}$(dose), a sigmoid curve will generally be the result. The transformation is needed in order to obtain a straight line. If k dose levels are considered and the proportion of units responding at level i is p_i (i = 1,..., k), a plot of a suitable transformation t(p) of the p_i against $\log_{10}$(dose) will be an approximate straight line. The line can be fitted by the usual least-squares method. However, with quantal data, the points do not all have equal weights, even though the number of experimental units in each group is the same. This is the main difference between the case of quantal response and that of a quantitative response.

4.4. *Linear Transformations for Sigmoid Curves: Tolerance Distribution*

When the response is quantal (or binary) its occurrence or non-occurrence will depend upon the intensity of the stimulus administered. Tolerance is defined as the level of intensity below which the response does not occur and above which the response occurs. The tolerance varies from unit to unit and thus one can talk of the distribution of tolerances. If f(x) denotes the density of the tolerance distribution, then the proportion of people responding to a dose x_0 is given by P where

$$P = \int_0^{x_0} f(x)dx,$$

or P can be interpreted as the probability that the unit chosen at random will respond to the stimulus at dose level x_0. Then one can look upon P, purely as a function of x_0 satisfying certain postulates. Since P is zero for small x (and unity for large x and is strictly increasing in x), it acts like a distribution function. Then the models that are proposed in the literature are:

(1) Normal curve:

$$P = (2\pi)^{-1/2} \int_{-\infty}^{\alpha+\beta x} \exp(-u^2/2) du, \quad -\infty < x < \infty,$$

probit P = normal deviate = $\alpha + \beta x$.

Note: The term 'probit' was used by Bliss [1934] as an abbreviation for 'probability unit'.

(2) Logistic curve:

$$P = [1 + \exp(-\alpha - \beta x)]^{-1}, \quad -\infty < x < \infty.$$

The logit P is given by

$$\text{logit } P = \ln \frac{P}{1-P} = \alpha + \beta x.$$

The term 'logit' was used by Berkson which is an abbreviation of logistic unit. Some people call $\ln[P/(1-P)]$ as log odds. Fisher and Yates [1963] define logit P as $(1/2)\ln[P/(1-P)]$.

(3) Sine curve:

$$P = (1/2)[1 + \sin(\alpha + \beta x)], \quad -\pi/2 \leq \alpha + \beta x \leq \pi/2.$$

This curve, proposed by Knudson and Curtis [1947], allows for finite range of doses. One can have the following angular transformation:

$$k = \sin^{-1}(2P - 1) = \alpha + \beta x.$$

The form given in Fisher and Yates [1963] is $k = \sin^{-1}(P^{1/2})$.

(4) Urban's curve:

$$P = 1/2 + (1/\pi)\tan^{-1}(\alpha + \beta x), \quad -\infty < x < \infty.$$

This gives

$$\tan[(2P - 1)\pi/2] = \alpha + \beta x.$$

Since $dP = 1/\pi[1 + (\alpha + \beta x)^2]$, it is the Cauchy tolerance distribution. The last three curves have the general form:

$$P = (1/2)[1 + F(\alpha + \beta x)].$$

For the logistic

$F(\alpha + \beta x) = \tanh[(\alpha + \beta x)/2]$.

If the tails are ignored, the models in (1)–(4) look alike. For comparing the four models we set $\alpha = 0$ since it denotes the origin of the scale. Also note that the curves are skew symmetric, that is $P(-x) = 1 - P(x)$, about the point $x = 0$, $P = \frac{1}{2}$.

Other transformations that are also used are

$P = 1 - e^{-\beta t}$ so that $-\ln(1 - P) = \beta t$.

If $-\ln(1 - P)$ is plotted against t, then one should get an approximate straight line. The log transformation is given by

$\ln[-\ln(1 - P)] = \log \beta + \log t$.

This will be useful if one is interested in estimating β_1/β_2 rather than a single β.

4.5. *Importance and Properties of the Logistic Curve*

The logistic curve has been used as a model for growth. It is a special case of a general function considered by Richards [1959]. If

$y = [1 + \exp(-\alpha - \beta x)]^{-1}$

then it satisfies the differential equation

$dy/dx = \beta y(1 - y)$

which enables one to give a more meaningful interpretation of the underlying phenomena. If y denotes the 'mass' or 'size' of some quantity, then the rate of change of this mass is proportional to the mass and a factor which decreases as the mass increases. Typically x denotes the time.

Properties of the Logistic Curve

The function y given above has all the properties of a cumulative distribution function; $x = 0$ is a point of inflection because $d^2y/dx^2 = \beta(1 - 2y)(dy/dx)$ vanishes at $x = 0$; dy/dx achieves its maximum $\beta/4$ at $x = 0$. The expected value of dy/dx is

$\int_0^1 \beta y(1 - y) dy = \beta/6$

and is called the mean growth rate. Thus $6/\beta$ is called the average time required for the major part of the growth to be achieved.

4.6. *Estimation of the Parameters*

If

$$P = [1 + \exp(-\alpha - \beta x)]^{-1}$$

then the logit denoted by l is

$l = \text{logit } P = \ln[P/(1 - P)] = \alpha + \beta x.$

Special graph papers are available for fitting a straight line to the logit data. Pearl [1924], Schultz [1930] and Davis [1941] provide some simple methods of estimating the parameters α and β. Oliver [1964] gives a review of the above three early methods of estimation.
Also by writing $\exp(-\alpha - \beta x) = \exp[-\beta(x - \mu)]$ where $\mu = -\alpha/\beta$, one infers that μ denotes the median effective dose or the median lethal dose denoted by LD_{50} or ED_{50}. Its estimate is $c = -a/b$. c is somewhat insensitive to the choice of origin on the x-axis and is invariant under scale changes.

Method of Maximum Likelihood

Berkson [1957b] extended Fisher's [1935] iterative method of maximum likelihood to estimate the parameters of the logistic curve. Let R_i denote the number of experimental units responding to the i-th dosage level, namely x_i. Then

$$P(R_i = r_i | x_i) = \binom{n_i}{r_i} P_i^{r_i} Q_i^{n_i - r_i} = C_i P_i^{r_i} Q_i^{n_i - r_i}, \quad Q_i = 1 - P_i,$$

where n_i denote the number of experimental units given the dosage x_i (i = 1,..., k). Then the log likelihood of the parameters is

$L = \sum \ln C_i + \sum r_i \ln P_i + \sum (n_i - r_i) \ln Q_i,$ since

$$P(x) = [1 + \exp(-\alpha - \beta x)]^{-1}$$

note that $\partial P/\partial \alpha = PQ$ and $\partial P/\partial \beta = xPQ$.
Hence

$$\frac{\partial L}{\partial \alpha} = \sum_{i=1}^{k} \frac{r_i}{P_i} \frac{\partial P_i}{\partial \alpha} - \sum \left(\frac{n_i - r_i}{Q_i}\right) \frac{\partial P_i}{\partial \alpha}$$

$$= \sum_{i=1}^{k} \frac{\partial P_i}{\partial \alpha} \left(\frac{r_i - n_i P_i}{P_i Q_i}\right) = \sum (r_i - n_i P_i) \quad \text{and}$$

$$\frac{\partial L}{\partial \beta} = \frac{\partial P_i}{\partial \beta} \left[\frac{(r_i - n_i P_i)}{P_i Q_i}\right] = \sum x_i (r_i - n_i P_i).$$

The Logit Approach

Let $r_i/n_i = p_i = i = 1, \ldots, k$. Then

$$\frac{\partial L}{\partial \alpha} = \sum n_i(p_i - P_i) \quad \text{and}$$

$$\partial L/\partial \beta = \sum n_i x_i(p_i - P_i).$$

Also since

$$L = \sum \ln C_i + \sum n_i \ln Q_i + \sum r_i \ln(P_i/Q_i), \quad \text{where } \ln(P_i/Q_i) = (\alpha + \beta x_i),$$

it is easy to note that $\sum r_i$ and $\sum r_i x_i$ are jointly minimally sufficient for (α, β). Berkson [1957b] has suggested an iterative method of determining α and β from the likelihood equations.

An estimate of the logit l_i is $\hat{l}_i = \ln[r_i/(n_i - r_i)]$ $(i = 1, \ldots, k)$. By plotting $(\hat{l}_i, x_i)$ $i = 1, \ldots, k$, let a_0 and b_0 be the preliminary estimates of α and β obtained by fitting a line by eye to the observations. Then let $\hat{l}_{0i} = a_0 + b_0 x_i$ be the provisional value of the logit $\hat{l}_i$, and $\hat{P}_{0i} = \{1 + \exp(-\hat{l}_{0i})\}^{-1}$ denote the corresponding value for $\hat{P}_i$. Also since

$$dl = dP/PQ$$

we can approximately take

$$\hat{P}_{0i} - \hat{P}_i = (\hat{l}_{0i} - \hat{l}_i)\hat{P}_{0i}\hat{Q}_{0i} = -(\delta a + x_i \delta b)\hat{P}_{0i}\hat{Q}_{0i}.$$

Substitution of this formula in the least-squares equations, namely $\partial L/\partial \alpha = 0$ and $\partial L/\partial \beta = 0$, we obtain (after writing $p_i - \hat{P}_i = p_i - \hat{P}_{i0} + \hat{P}_{i0} - \hat{P}_i$)

$$\sum n_i[(p_i - \hat{P}_{i0}) - (\hat{l}_i - \hat{l}_{0i})\hat{P}_{0i}\hat{Q}_{0i}] = 0$$

$$\sum n_i x_i[(p_i - \hat{P}_{i0}) - (\hat{l}_i - \hat{l}_{0i})\hat{P}_{0i}\hat{Q}_{0i}] = 0.$$

Hence

$$\delta a(\sum n_i \hat{P}_{0i}\hat{Q}_{0i}) + \delta b(\sum n_i \hat{P}_{0i}\hat{Q}_{0i} x_i) = \sum n_i(p_i - \hat{P}_{0i}),$$

$$\delta a(\sum n_i \hat{P}_{0i}\hat{Q}_{0i} x_i) + \delta b(\sum n_i \hat{P}_{0i}\hat{Q}_{0i} x_i^2) = \sum n_i x_i(p_i - \hat{P}_{0i}).$$

Alternatively, one can write these equations as

$$\begin{vmatrix} \partial^2 L/\partial a_0^2 & \partial^2 L/\partial a_0 \partial b_0 \\ \partial^2 L/\partial a_0 \partial b_0 & \partial^2 L/\partial b_0^2 \end{vmatrix} \begin{vmatrix} \delta a \\ \delta b \end{vmatrix} = - \begin{vmatrix} \partial L/\partial a_0 \\ \partial L/\partial b_0 \end{vmatrix}$$

where the subscripts on the derivatives denote that after differentiation they are evaluated at a_0 and b_0.

$$\begin{vmatrix} \delta a \\ \delta b \end{vmatrix} = \begin{vmatrix} \sum n_i x_i^2 P_{0i} Q_{0i} & -\sum n_i x_i P_{0i} Q_{0i} \\ -\sum n_i x_i P_{0i} Q_{0i} & \sum n_i P_{0i} Q_{0i} \end{vmatrix} \begin{vmatrix} \partial L/\partial a_0 \\ \partial L/\partial b_0 \end{vmatrix} / D$$

where $D = (\sum n_i P_{0i} Q_{0i})(\sum n_i x_i^2 P_{0i} Q_{0i}) - (\sum n_i P_{0i} Q_{0i} x_i)^2$. After evaluating δa and δb, one can go to the next iteration, namely

$a_2 = a_1 + \delta a$, $b_2 = b_1 + \delta b$ where $a_1 = a_0 + \delta a$, $b_1 = b_0 + \delta b$.

This process of iteration is repeated until a certain internal consistency is obtained. Tables providing antilogits P_i and weights $W_i = P_i Q_i$ for various values of l_i are provided by Berkson [1953].

Notice that the procedure requires initial starting values. Hodges [1958] suggests a method called the 'transfer method' which enables one to obtain reasonable starting values. The method is based on minimal sufficient statistics of $\sum r_i$ and $\sum r_i x_i$. The sets $\{r_i\}$ and $\{r'_i\}$ are said to be equivalent if $\sum r_i = \sum r'_i$ and $\sum x_i r_i = \sum x_i r'_i$. Now choose the set $\{r'_i\}$ such that the points (x_i, l'_i), where l'_i is the logit corresponding to r'_i, are on the same line. Then the mles of α and β can be obtained from the line $l' = a + bx$ drawn through these points. The problem is how to construct the set $\{r'_i\}$. Now $\sum r'_i = \sum r_i$ implies that r_i can be interchanged among the levels of the doses, however, $\sum x_i r_i = \sum x_i r'_i$ implies that the interchanges are subjected to the restriction that the line passes through the center of gravity of the responses. If the original data are plotted on logit paper and a straight line is drawn by eye, certain points fall below and some others fall above the line. A method of transfer of points can be designed in order to achieve collinearity of the points. r_i can be adjusted to produce r'_i and another line can be fitted. After 3 or 4 transfer of points, satisfactory collinearity can be achieved. This procedure is simple to carry out when the x_i are equally spaced.

Variances of the mle Values of α and β

Let $\bar{x} = \sum n_i W_i x_i / \sum n_i W_i$, $W_i = P_i Q_i$ and $\hat{l} = a + bx = a' + b(x - \bar{x})$, $a' = a + b\bar{x}$. The large-sample variances of a and b can be obtained from the inverse of $\sum$ where

$$\sum = \begin{vmatrix} \sum n_i W_i & \sum n_i W_i x_i \\ \sum n_i W_i x_i & \sum n_i W_i x_i^2 \end{vmatrix}$$

$\det \sum = \{[\sum (n_i W_i)][\sum n_i W_i (x_i - \bar{x})^2]\}$, $\bar{x} = \sum n_i W_i x_i / \sum n_i W_i$, and

$$\sum^{-1} = (\det \sum)^{-1} \begin{vmatrix} \sum n_i W_i x_i^2 & -\sum n_i W_i x_i \\ -\sum n_i W_i x_i & \sum n_i W_i \end{vmatrix}.$$

Thus

$\sigma_a^2 = \sum n_i W_i x_i^2 / (\det \sum) = (\sum n_i W_i)^{-1} + \bar{x}^2 [\sum n_i W_i (x_i - \bar{x})^2]^{-1}$

$\sigma_b^2 = [\sum n_i W_i (x_i - \bar{x})^2]^{-1}$

$\text{cov}(a,b) = -\bar{x} / \sum n_i W_i (x_i - \bar{x})^2$.

The Logit Approach

Hence

$$\sigma_{a'}^2 = \sigma_a^2 - \bar{x}^2 \sigma_b^2 = 1/\sum n_i W_i.$$

If X/Y is designed to estimate θ_1/θ_2, then

$$E(X/Y) \approx \theta_1/\theta_2.$$

One can show by the differential approach that (see section 2.2)

$$\text{var}(X/Y) = (\text{var}\, X)\theta_2^{-2} + \theta_1^2(\text{var}\, Y)\theta_2^{-4} - 2\theta_1 \text{cov}(X,Y)\theta_2^{-3}.$$

So if $c = -a/b$ is an estimate of $\mu = -\alpha/\beta$, then $\sigma_c^2 = \text{var}(-a/b) = \text{var}(a/b)$ and hence

$$s_c^2 = b^{-2}[\sigma_a^2 + a^2 b^{-2}\sigma_b^2 - 2ab^{-1}\text{cov}(a,b)] = b^{-2}[\sigma_a^2 + (c^2 - 2c\bar{x})\sigma_b^2]$$

$$= b^{-2}[(\sum n_i W_i)^{-1} + (c - \bar{x})^2/\sum n_i W_i(x_i - \bar{x})^2]$$

$$= b^{-2}[\sigma_{a'}^2 + (c - \bar{x})^2 \sigma_b^2].$$

The assumptions governing the validity of the above expressions for the large-sample expressions for the variances of a, b and c are:
(1) the true probabilities conform to the logistic function,
(2) the n_i are large, and
(3) the dose levels are free of measurement errors.
Hence, the standard errors of a, b and c should be treated as approximations.

4.7. *Estimation of the Parameters in the Probit by the Method of Maximum Likelihood*

Let $P_i = \Phi(\alpha + \beta x_i)$, where Φ denotes the standard normal distribution function and α, β are the unknown parameters and x_i is the dose level. Let n_i be the number of experimental units at dose level x_i and r_i denote the number responding to the dose ($i = 1, \ldots, k$). Then the likelihood of α and β is

$$L = \sum \ln C_i + \sum r_i \ln P_i + \sum (n_i - r_i)\ln(1 - P_i), \quad \text{where } C_i = \binom{n_i}{r_i}$$

$$\frac{\partial L}{\partial \alpha} = \sum n_i \frac{(p_i - P_i)}{P_i Q_i} \frac{\partial P_i}{\partial \alpha}, \quad p_i = r_i/n_i$$

$$\frac{\partial L}{\partial \beta} = \sum n_i \frac{(p_i - P_i)}{P_i Q_i} \frac{\partial P_i}{\partial \beta}, \quad \text{where}$$

$$\frac{\partial P_i}{\partial \alpha} = \phi(\alpha + \beta x_i), \quad \frac{\partial P_i}{\partial \beta} = x_i \phi(\alpha + \beta x_i).$$

Let us compute the information matrix

$$E\frac{\partial^2 L}{\partial \alpha^2} = -\sum n_i \frac{1}{P_i Q_i}\left(\frac{\partial P_i}{\partial \alpha}\right)^2 = -\sum n_i \phi^2(\alpha + \beta x_i)/P_i Q_i$$

$$E\frac{\partial^2 L}{\partial \alpha \partial \beta} = -\sum \frac{n_i}{P_i Q_i}\frac{\partial P_i}{\partial \alpha}\frac{\partial P_i}{\partial \beta} = -\sum n_i x_i \phi_i^2/P_i Q_i$$

$$E\frac{\partial^2 L}{\partial \beta^2} = -\sum n_i x_i^2 \phi_i^2/P_i Q_i, \quad \phi_i = \phi(\alpha + \beta x_i).$$

Let $W_i = \phi_i^2/P_i Q_i$. Then the information matrix is

$$\begin{vmatrix} \sum n_i W_i & \sum n_i x_i W_i \\ \sum n_i x_i W_i & \sum n_i W_i x_i^2 \end{vmatrix}$$

Consider

$$\frac{\partial L}{\partial \alpha} = \sum n_i \frac{(p_i - P_i)}{P_i Q_i}\phi(\alpha + \beta x_i) = 0.$$

Let

$$H_i(\alpha, \beta) = \frac{p_i - P_i}{P_i Q_i}\phi(\alpha + \beta x_i).$$

Let a_0 and b_0 be the initial values of a and b, the estimates of α and β. Then

$$H_i(a,b) = H_i(a_0, b_0) + (a - a_0)\frac{\partial H_{i,0}}{\partial a_0} + (b - b_0)\frac{\partial H_{i0}}{\partial b_0}$$

$$\frac{\partial H}{\partial a} = \frac{-\partial P}{\partial a}\frac{\phi(a+bx)}{PQ} - \frac{(p-P)(a+bx)\phi}{PQ} - \frac{(p-P)}{(PQ)^2}(1-2P)\phi^2$$

$$= \frac{-\phi^2}{PQ} - \frac{(p-P)(a+bx)\phi}{PQ} - \frac{(p-P)}{(PQ)^2}(1-2P)\phi^2.$$

Let $1 = \Phi^{-1}(P) = G(P) = \alpha + \beta x$. Then

$$\frac{\partial P}{\partial \alpha} = \frac{\partial P}{\partial 1}\cdot\frac{\partial 1}{\partial \alpha} = \frac{\partial P}{\partial 1},$$

$$\frac{\partial 1}{\partial P} = G'(P) = \frac{1}{\phi[\Phi^{-1}(P)]} = \frac{1}{\phi(1)}, \quad \text{and}$$

$$(p-P)\frac{dP}{d\alpha} = -[\Phi(1) - \Phi(\hat{1})]\frac{dP}{d\alpha} \doteq -(1-\hat{1})\phi^2(\hat{1}).$$

Likelihood equations are

$$\sum n_i \left(\frac{1_i - \hat{1}_i}{P_i Q_i}\right)\phi^2(\hat{1}_i) = 0,$$

$$\sum n_i x_i \left(\frac{1_i - \hat{1}_i}{P_i Q_i}\right)\phi^2(\hat{1}_i) = 0.$$

Now writing

$$l_i - \hat{l}_i = l_i - \hat{l}_{i0} + \hat{l}_{i0} - \hat{l}_i, \quad \hat{l}_{i0} = a_0 + b_0 x_i$$
$$= l_i - \hat{l}_{i0} - (\delta a_0 + \delta b_0 x_i)$$

we have

$$\sum n_i W_{i0}(\delta a_0 + \delta b_0 x_i) = \sum n_i W_{i0}(l_i - \hat{l}_{i0})$$
$$\sum n_i x_i W_{i0}(\delta a_0 + \delta b_0 x_i) = \sum n_i x_i W_{i0}(l_i - \hat{l}_{i0}).$$

That is

$$\sum n_i W_{i0}(a_1 + b_1 x_i) = \sum n_i W_{i0} l_i$$
$$\sum n_i x_i W_{i0}(a_1 + b_1 x_i) = \sum n_i x_i W_{i0} l_i, \quad \text{where}$$
$$W_{i0} = \phi^2(\hat{l}_{i0})/\hat{P}_{i0} \hat{Q}_{i0}.$$

Thus

$$\begin{vmatrix} a_1 \\ b_1 \end{vmatrix} = \begin{vmatrix} \sum n_i W_{i0} & \sum n_i x_i W_{i0} \\ \sum n_i x_i W_{i0} & \sum n_i x_i^2 W_{i0} \end{vmatrix}^{-1} \begin{vmatrix} \sum n_i W_{i0} l_i \\ \sum n_i x_i W_{i0} l_i \end{vmatrix}.$$

The iteration is carried out with $l_i = \Phi^{-1}(p_i)$ until the estimates do not differ significantly.

Remark. Adjustment for natural mortality rate can also be made. If c denotes the proportion of the population that will respond even to zero dose, then the probability that an experimental subject will respond at dose level x is

$$P^*(x) = 1 - (1 - P)(1 - c) = c + (1 - c)P(x).$$

Here, c is called the natural mortality rate. This is known as Abbott's formula. After obtaining an estimate of c from a controlled group the method of maximum likelihood can be carried out by replacing $P^*(x)$ by $c + (1 - c)P(x)$ in the likelihood equations.

4.8. Other Available Methods

Even though response is a continuous variable we may classify the observations into mutually exclusive categories or the data might naturally arise as classified into mutually exclusive categories. Let n denote the total number of observations, n_i be the number of observations falling in the i-th category and $\pi_i(\theta)$ denote the probability of our observation falling in the i-th category ($i = 1, \ldots, k$). Notice that θ, the unknown parameter, could be vector-valued. In the following we shall give some well-known methods of estimation and the criterion employed.

(1) Minimum χ^2 method. Find the θ which minimizes

$$\sum_{i=1}^{k} [n_i - n\pi_i(\theta)]^2 / n\pi_i(\theta) = \sum_{1}^{k} \frac{n_i^2}{n\pi_i(\theta)} - n.$$

(2) Minimum modified χ^2. Find the θ that minimizes

$$\sum [n_i - n\pi_i(\theta)]^2/n_i = -n + \sum_1^k n^2 \pi_i(\theta)/n_i.$$

with n_i replaced by unity if $n_i = 0$.

(3) Hellinger Distance (HD). Find θ that minimizes

$$HD = \cos^{-1}[\sum (n_i/n)\pi_i(\theta)]^{1/2}$$

(4) Kullback-Leibler Separator (KLS)

$$KLS = \sum \pi_i(\theta) \log[\pi_i(\theta)/(n_i/n)]$$

(5) Haldane's Discrepancy

$$D_r = \frac{(n+r)!}{n!} \sum \frac{n_i! \pi_i^{r+1}(\theta)}{(n_i+r)!} \quad (r \neq -1)$$

$$D_{-1} = -n^{-1} \sum n_i \log \pi_i(\theta).$$

Rao [1965, p. 289] points out that the maximum likelihood method is superior to any of the preceding ones, from the point of second-order efficiency. Under certain regularity conditions, the maximum likelihood estimators of $\underline{\theta}$ are asymptotically normal $(\underline{\theta}, \sum)$ where $\sum^{-1}$ denotes the Fisher information matrix.

4.9. *Method of Minimum Logit χ^2*

We wish to find α and β that would minimize $\sum W_i'(p_i - P_i)^2$, where $W_i' = n_i/P_i Q_i$ = reciprocal of the variance of p_i. The above expression is equivalent to Pearson's χ^2 test statistic because

$$\chi^2 = \sum_i (Q_i - E_i)^2/E_i$$

$$= \sum \frac{\left(\frac{r_i}{n_i} - P_i\right)^2}{P_i} + \frac{\left(1 - \frac{r_i}{n_i} - Q_i\right)^2}{Q_i} n_i$$

$$= \sum n_i (p_i - P_i)^2/P_i Q_i.$$

Let $\hat{l}_i = \ln(p_i/q_i)$, then one can write

$p_i - P_i \doteq P_i Q_i (l_i - \hat{l}_i)$ (for the logistic case only) or

$(p_i - P_i)^2 \doteq P_i Q_i (p_i q_i)(l_i - \hat{l}_i)^2.$

Hence

$$\chi^2 \doteq \sum n_i p_i q_i (l_i - \hat{l}_i)^2.$$

One can easily minimize the above expression without iterations. This property is unique to the logistic case and is not shared by the cumulative normal curve. The method of estimation is the method of weighted least squares. Setting $W_i = p_i q_i$, the equations are

$$\sum n_i W_i(a + bx_i) = \sum n_i W_i l_i$$

$$\sum n_i x_i W_i(a + bx_i) = \sum n_i W_i x_i l_i, \quad l_i = \ln(p_i/q_i) = \ln(r_i/(n_i - r_i)).$$

Alternatively,

$$\begin{vmatrix} \sum n_i W_i & \sum n_i W_i x_i \\ \sum n_i W_i x_i & \sum n_i W_i x_i^2 \end{vmatrix} \begin{vmatrix} a \\ b \end{vmatrix} = \begin{vmatrix} \sum n_i W_i l_i \\ \sum n_i W_i x_i l_i \end{vmatrix}.$$

Hence

$$\begin{vmatrix} a \\ b \end{vmatrix} = |S|^{-1} \begin{vmatrix} \sum n_i W_i l_i \\ \sum n_i W_i x_i l_i \end{vmatrix}$$

where

$$|S|^{-1} = \begin{vmatrix} \sum n_i W_i x_i^2 & -\sum n_i W_i x_i \\ -\sum n_i W_i x_i & \sum n_i W_i \end{vmatrix} / D$$

$$D = (\sum n_i W_i)[\sum n_i W_i(x_i - \bar{x})^2].$$

$\bar{x} = \sum n_i W_i x_i / \sum n_i W_i$ and $\bar{l}$ is defined analogously. So,

$$b = \frac{(\sum n_i W_i)(\sum n_i W_i x_i l_i) - (\sum n_i W_i l_i)(\sum n_i W_i x_i)}{D}$$

$$= \sum n_i W_i (l_i - \bar{l})(x_i - \bar{x}) / \sum n_i W_i (x_i - \bar{x})^2$$

$$a = \bar{l} - b\bar{x} = (\sum n_i W_i l_i - b \sum n_i W_i x_i)/\sum n_i W_i.$$

Anscombe [1956] points out that l is biased as an estimate of $\alpha + \beta x$. In order to remove the bias almost completely, he proposes to refine l to

$$l(r) = l = \ln[(r + \tfrac{1}{2})/(n - r + \tfrac{1}{2})].$$

Now expanding l in Taylor series about $r = nP$ and taking expectations, one can show that $l(nP) = \ln(P/Q) + (Q - P)/2nPQ + O(n^{-2})$, $l'(nP) = (nPQ)^{-1} - (P^{-2} + Q^{-2})/2n^2 + O(n^{-3})$ and $l''(nP) = -(nP)^{-2} + (nQ)^{-2} + O(n^{-3})$. Hence

$$El = \alpha + \beta x + O(n^{-2}), \quad \text{var} \, l = (nPQ)^{-1} + O(n^{-2})$$

and the skewness of $l \simeq 2(P - Q)/(nPQ)^{1/2}$. Notice that the skewness in l is twice that of r and is opposite in sign.

After considering the size and sign of the bias for various situations, Anscombe [1956] advocates the modified definition of l and the minimum χ^2 method, with the weight W, defined empirically by $W = r(n - r)/n^2$, or W is replaced by a fitted weight $\hat{W} = \hat{P}\hat{Q}$ if $n\hat{W}$ exceeds 1 and equal to 0 otherwise. Tukey (see Anscombe [1956]) suggested a further improvement, especially for small n, namely setting

$$l = \ln\left(\frac{r + \frac{1}{2}}{n - r + \frac{1}{2}}\right) + \frac{1}{2} \quad \text{when } r = n$$

$$= \ln\left(\frac{r + \frac{1}{2}}{n - r + \frac{1}{2}}\right) - \frac{1}{2} \quad \text{when } r = 0.$$

so that the range of values of l is widened a little.

One might ask whether one should prefer the minimum logit method in comparison with the method of maximum likelihood. The minimum χ^2 method does not need the elaborate iterations whereas the method of maximum likelihood does. The latter produces estimates that are functions of the sufficient statistic ($\sum r_i, \sum r_i x_i$). However, the method of minimum χ^2 depends only on $(r_1, \ldots, r_k)$. Taylor [1953] has shown that the minimum logit χ^2 estimates are RBAN (regular best asymptotically normal) and efficient and hence are asymptotically equivalent to the mle values. A draw-back of the minimum logit method is that when the number of experimental units at the dose levels are small, then the r_i will be small and hence the value of χ^2 will be unstable.

Little [1968] has shown algebraically that the minimum logit method yields estimates for the slope that are consistently smaller (in absolute magnitude) than the mle estimates. When the slope is positive, the intercept is consistently larger than that yielded by the method of maximum likelihood. Berkson [1955, 1968] addresses himself to these issues and provides answers.

4.10. *Goodness-of-Fit Tests*

The minimum logit method is based on the assumption that χ^2 is not significant. However, it is desirable to test the goodness of the model before we estimate the parameters. We can use either Pearson's χ^2 or the minimum logit χ^2 with the appropriate number of degrees of freedom in order to test the goodness of the model, given by

Pearson $\chi^2 = \sum n_i (p_i - \hat{P}_i)^2 / \hat{P}_i \hat{Q}_i$

logit $\chi^2 = \sum n_i p_i q_i (l_i - \hat{l}_i)^2$.

However, Anscombe [1956] points out that the above method of goodness-of-fit will be unsatisfactory since χ^2 will be sensitive to departures from the binomial distribution of responses at any dose level. A better procedure would be to estimate β_2 and β_3 which are assumed to be small where

$$\ln(P/Q) = \sum_{j=1}^{3} \beta_j (x - \mu)^j.$$

The appropriate statistics to be used for estimating these parameters are $\sum_{j=1}^{k} r_i x_i^j$ ($j = 0, 1, 2, 3$). We need large series of observations at each dose level and we assume that the dose levels cover a wide range. If $\beta_2 \neq 0$, the response curve of P against x is not antisymmetrical about $x = \mu$. This may be remedied by a suitable transformation of the x scale. If $\beta_2 = 0$ and $\beta_3 \neq 0$, the logistic curve is not suitable and one should look at other curves.

4.11. *Spearman-Karber Estimator*

If in section 4.6 the x values are equally and not too widely spaced, such that PQ may be assumed to be practically equal to zero outside the range of x values, and all the n_i values are equal, then $\sum P_i$ and $\sum x_i P_i$ can be replaced by the appropriate integrals using the Euler-MacLaurin formula. For instance, if x_i is the i-th dose level, h is the spacing, and r_i out of n_i respond at dose level x_i ($i = 1, \ldots, k$), let

$$P = [1 + e^{-\beta(x-\mu)}]^{-1}, \quad \mu = -\alpha/\beta.$$

Then

$$\mu = LD_{50} = \int_0^{x_k} x dP = \sum_{i=1}^{k} \int_{x_{i-1}+h/2}^{x_i+h/2} x dP$$

$$\doteq \sum_{i=1}^{k} [(x_i + h/2)P_i - (x_{i-1} + h/2)P_{i-1} - hP_i]$$

$$= x_k + h/2 - h \sum_{i=1}^{k} P_i, \quad \text{since } P_k = 1$$

after performing partial integration once. Now, the estimate of μ is

$$\hat{\mu} = x_k + h/2 - h \sum_{i=1}^{k} p_i \quad (p_i = r_i/n_i, i = 1, \ldots, k)$$

which is the Spearman-Karber formula for ED_{50}. Since the second moment of the logistic distribution is $\mu^2 + \pi^2/3\beta^2$, we have

$$\mu^2 + \pi^2/3\beta^2 = \int x^2 dP \doteq \sum_{i=1}^{k} \int_{x_{i-1}+h/2}^{x_i+h/2} x^2 dP$$

$$= \sum [(x_i + h/2)^2 P_i - (x_{i-1} + h/2)^2 P_i - 2P_i x_i h]$$

$$= (x_k + h/2)^2 - 2h \sum x_i P_i, \quad \text{since } P_k = 1.$$

An estimate of β can be obtained from the estimate of the right side quantity given by

$$(x_k + h/2)^2 - 2h \sum x_i p_i.$$

After solving, we obtain

$$\hat{\beta} = (3/\pi^2)^{1/2} [h(2x_k + h) \sum p_i - h^2 (\sum p_i)^2 - 2h(\sum x_i p_i)]^{-1/2}.$$

Notice that the estimate for β was surmised by Anscombe [1956]. Also Anscombe [1956] obtains $-h^2/12$ in addition to the expression for $\mu^2 + \pi^2/3\beta^2$. This can be attributed to 'Sheppard's correction'.

The Spearman-Karber estimator is nonparametric in the sense that it can be computed without specifying the functional form of the dose-response function. If x_k does not denote the largest dose level (i.e. when $P_k \neq 1$), Spearman [1908] proposed the following estimator for LD_{50} by

$$\bar{x} = p_1(x_1 - h/2) + \sum_{i=1}^{k-1}(x_i + h/2)(p_{i+1} - p_i) + (1 - p_k)(x_k + h/2)$$

$$= \sum_{j=2}^{k}(x_j - h/2)p_j - \sum_{i=2}^{k}(x_i + h/2)p_i - (x_1 + h/2)p_1 + p_1(x_1 - h/2) + (x_k + h/2)$$

$$= \sum_{j=2}^{k}(x_{j-1} - x_j)p_j - hp_1 + x_k + h/2 = x_k + h/2 - h\sum_{j=1}^{k}p_j.$$

4.11.1. *The Infinite Experiment*

Consider an experiment having infinitely many dose levels given by $x_i = x_0 + ih$ ($i = 0, \pm 1, \pm 2, \ldots$). Assume that the dose-mesh location x_0 and the dose interval h will be arbitrarily chosen and fixed for the time being. The variance of r_i at dose level x_i is $n_i P_i(1 - P_i)$. Hence, the variance at all but a finite number of dose levels is negligible. Thus, results pertaining to infinite experiments will be relevant to finite experiments with dose levels covering the range of variance. Brown [1961] has studied certain properties of the Spearman estimator for the infinite experiment given by

$$\bar{x} = \sum_{i=-\infty}^{\infty}(x_i + h/2)(p_{i+1} - p_i)$$

or

$$\bar{x} = x_0 + h/2 + h\sum_{i=1}^{\infty}(1 - p_i) - h\sum_{i=-\infty}^{0}p_i$$

because

$$\sum_{i=-\infty}^{0}i(p_{i+1} - p_i) = -\sum_{i=-\infty}^{0}p_i \quad \text{and} \quad \sum_{i=1}^{\infty}i(p_{i+1} - p_i) = \sum_{i=1}^{\infty}iq_i = \sum_{i=1}^{\infty}\sum_{j=1}^{i}q_i = \sum_{j=1}^{\infty}\sum_{i=j}^{\infty}q_i = \sum_{j=1}^{\infty}(1 - p_j).$$

These will be presented in the following. If $P(x)$ has a finite mean μ, then the series defining $\bar{x}$ converges with probability 1 and

$$E(\bar{x}|x_0) = \sum_{i=-\infty}^{\infty}(x_i + h/2)(P_{i+1} - P_i).$$

Lemma 1

$$|B(\bar{x}|x_0)| \leq h/2, \quad B(\bar{x}|x_0) = E(\bar{x}|x_0) - \mu.$$

Proof.

$$B(\bar{x}|x_0) = E(\bar{x}|x_0) - \int_{-\infty}^{\infty} x\,dP$$

$$= \sum_{i=-\infty}^{\infty} (x_i + h/2)(P_{i+1} - P_i) - \sum_{i=-\infty}^{\infty} \int_{x_i}^{x_{i+1}} x\,dP$$

$$= \sum_{i=-\infty}^{\infty} (x_i + h/2 - c_i)(P_{i+1} - P_i)$$

where

$$c_i = \int_{x_i}^{x_{i+1}} x\,dP \bigg/ \int_{x_i}^{x_{i+1}} dP \quad \text{and} \quad x_i \leq c_i \leq x_{i+1}.$$

Now the proof is complete since $|x_i + h/2 - c_i| \leq h/2$.

For a one-point distribution P, with mass point located at one of the dose levels, the mean will be this dose level, whereas the Spearman estimator will be h/2 units lower than that dose level. Thus, the Spearman estimator achieves the bound h/2.

Lemma 2

If $P(x)$ is differentiable and $P'(x)$ is unimodal, with $m = \max P'(x)$, then

$$|B(\bar{x}|x_0)| \leq mh^2/8.$$

Proof. See Brown [1961, pp. 296–297].
The bound on the bias of the Spearman estimator given by Lemma 2 is sharp since equality is achieved when $P'(x)$ is uniform and for suitable choices of x_0 and h.

4.11.2. *Variance of the Spearman Estimator*

If $n_1 = n_2 = \ldots = n$, then the variance of $\bar{x}$ is

$$V(\bar{x}|x_0) = (h^2/n) \sum_{-\infty}^{\infty} P_i(1 - P_i);$$

let the approximating integral be

$$\bar{V}(\bar{x}) = (h/n) \int_{-\infty}^{\infty} P(1 - P)\,dx.$$

Lemma 3

For all P, x_0, h and n

$$|V(\bar{x}|x_0) - \bar{V}(\bar{x})| \leq h^2/4n. \tag{1}$$

If P(x) is symmetric with two points of inflection, then

$$|V(\bar{x}|x_0) - \bar{V}(\bar{x})| \leq h^3 m/3n \quad \text{where } m = \max P'(x). \tag{2}$$

Proof for (1). Let m^* be a median of P(x). Then $P(1-P)$ is nondecreasing for $x \leq m^*$ and nonincreasing for $x \geq m^*$. Number the x_i values so that $x_0 \leq m^* \leq x_1$. Now

$$\int_{x_{i-1}}^{x_i} P(1-P)dx \leq P_i(1-P_i)h \quad \text{for } i = 0, -1, -2, \ldots$$

$$\int_{x_i}^{x_{i+1}} P(1-P)dx \leq P_i(1-P_i)h \quad \text{for } i = 1, 2, 3, \ldots$$

$$\int_{x_0}^{x_1} P(1-P)dx \leq \tfrac{1}{2} \cdot \tfrac{1}{2} \cdot h.$$

Combining the preceding inequalities we obtain

$$h \sum_{-\infty}^{\infty} P_i(1-P_i) + h/4 \geq \int_{-\infty}^{\infty} P(1-P)dx.$$

Multiplying both sides of the above inequality by h/n we have

$$V(\bar{x}|x_0) - \bar{V}(\bar{x}) \geq -h^2/4n.$$

By writing down inequalities going in the opposite direction, one can analogously show that the difference is bounded above by $h^2/4n$.

Remark. The inequality for the difference $V(\bar{x}|x_0) - \bar{V}(\bar{x})$ is sharp since equality is achieved for the two-point distribution with equal probabilities at the two points and with h and x_0 appropriately chosen. The computations of Brown [1961] yield $\bar{V} = 0.5642\rho\sigma^2/n$ and $0.5513\rho\sigma^2/n$ for the normal and logistic P(x) where σ^2 denotes the variance of the tolerance distribution and $h = \rho\sigma$.

Proof for (2). Inequality (2) can be obtained by using Euler-MacLaurin expressions for the variance $V(\bar{x}|x_0)$. If x_0 is random with a uniform distribution on $(0,h)$, then Brown [1961, section 6] shows that $E(\bar{x}) = \mu$ and $V(\bar{x}) = \bar{V}(\bar{x}) + O(h^2)$.

4.11.3. *Asymptotic Efficiency of the Spearman Estimator*

The size of the experiment can be measured by the number of subjects tested per unit interval on the dose scale. This is denoted by $n' = n/h$. Let n' increase and $n = 1$, that is $h = 1/n'$. Then either for fixed or random x_0

$$V(\bar{x}) = (n')^{-1} \int_{-\infty}^{\infty} P(1-P)dx + O(1/n'^2).$$

Let us denote the first term in $V(\bar{x})$ by $V_A(\bar{x})$, which can be called the asymptotic variance.

Brown [1961] extends Finney's [1950, 1952] concept of information for the finite experiment to the infinite experiment. The former considers infinite series of information terms corresponding to dose levels. The expectation of this series is taken with respect to the uniform (0,h) distribution of x_0. Brown [1961] obtains the information for the infinite experiment as

$$I = n' \int_{-\infty}^{\infty} [\partial/\partial \mu P(x)]^2 [P(x)(1 - P(x))]^{-1} dx.$$

Then the asymptotic efficiency E of the Spearman estimator is defined as the ratio of I^{-1} to the asymptotic variance $V_A(\bar{x})$. Thus

$$E = I^{-1}/V_A(\bar{x})$$

$$= \left[\int_{-\infty}^{\infty} P(x)[1 - P(x)] dx \int_{-\infty}^{\infty} [\partial/\partial \mu P(x)]^2 [P(x)(1 - P(x))]^{-1} dx \right]^{-1}.$$

$E = 0.9814$ when P is normal and equals 1.000 when p is logistic which coincides with the values of Finney [1950, 1952].
Also if

$$P_e(x) = K(e) \int_{-\infty}^{x} \left[1 + \frac{(x - \mu)^2}{1 + 2e} \right]^{-1-e} dt$$

then $E \to 0$ as $e \to 0$, since the asymptotic variance of $P_e(x)$ tends to infinity while the information is bounded. Notice that $P_e(x)$ tends to the Cauchy distribution as $e \to 0$.

The theoretical merit of the Spearman estimator has also been demonstrated by Miller [1972], and Church and Cobb [1973]. Chimiel [1976] proposed a Spearman type of estimator for the variance in the quantal response bioassay problem when the mean of the tolerance distribution is both known and unknown. The consistency, asymptotic normality and efficiency of the variance estimator are studied. A single nonparametric variance estimator was proposed also by Epstein and Churchman [1944]. They and Cornfield and Mantel [1950] give a few properties of this variance estimator. Chimiel [1976] also proposes Spearman-type percentile estimator and establishes similar asymptotic properties. The variance estimator when μ is known is defined by

$$v = \lim_{k \to \infty} \left[p_{-k}(x_{-k} - h/2 - \mu)^2 + \sum_{i=-k}^{k-1} (p_{i+1} - p_i)(x_i + h/2 - \mu)^2 + (1 - p_k)(x_k + h/2 - \mu)^2 \right]$$

$$= (x_0 + h/2 - \mu)^2 + 2h \sum_{i=-\infty}^{0} p_i(\mu - x_i) - 2h \sum_{i=1}^{\infty} (1 - p_i)(\mu - x_i).$$

One might ask for which class of distributions the asymptotic efficiency of the Spearman estimator is equal to 1. If $P(x) = F(x - \mu)$ where $F(x) = (1 + e^{-x})^{-1}$, namely the logistic distribution, and hence $f(x) = F'(x) = F(x)[1 - F(x)]$, the asymptotic efficiency equals 1. In the following we will show that for any other $F(x - \mu)$, differentiable everywhere and having mean μ, the asymptotic efficiency will be strictly less than 1. Toward this we need the following definitions.

If $P(x) = F(x - \mu)$, where $F(x)$ is differentiable everywhere and having a mean of 0, then the asymptotic efficiency of the Spearman estimator is

$$[e(F)]^{-1} = \left\{\int_{-\infty}^{\infty} F(x)[1 - F(x)]dx\right\}\left\{\int_{-\infty}^{\infty} f^2(x)\{F(x)[1 - F(x)]\}^{-1}dx\right\}.$$

Then we have the following theorem.

Theorem

$e(F) \leq 1$ for all F differentiable everywhere and having a mean of 0, with equality if and only if $F(x) = [1 + e^{-\beta x}]^{-1}$ for some $\beta > 0$.

Proof. One can write

$$[e(F)]^{-1} = E\left\{\frac{F(X)[1 - F(X)]}{f(X)}\right\} E\left\{\frac{f(X)}{F(X)[1 - F(X)]}\right\}.$$

Now applying Cauchy-Schwarz inequality we obtain

$$E\left\{\frac{f(X)}{F(X)[1 - F(X)]}\right\} \geq \left[E\left\{\frac{F(X)[1 - F(X)]}{f(X)}\right\}\right]^{-1}.$$

Hence $[e(F)]^{-1} \geq 1$ for all F which are differentiable everywhere with equality if and only if $f(x)/F(x)[1 - F(x)] = \beta$ (a constant). Integrating on both sides, we have $\ln[F/(1 - F)] = \beta x + \alpha$ or $F(x) = [1 + e^{-(\alpha + \beta x)}]^{-1}$. Now that $F(x)$ has zero expectation, $\alpha = 0$ is implied.

Remark. The notion of an infinite experiment is somewhat artificial. Infinite experiments can be treated as limits of finite experiments. For the alternative definition of asymptotic efficiency of the Spearman estimator and other results, the reader is referred to Govindarajulu and Lindqvist [1987].

Historical Note. For the finite experiment, Finney [1950, 1952] evaluated the asymptotic variances of the maximum likelihood estimator, averaged over the choices of x_0, for the normal and logistic tolerance distributions. He also considered the limit of the same as the number of levels tended to infinity. He compared these values with the mean square error of the Spearman estimator over choices of x_0 for the same two distributions. The ratios were 0.9814 and 1.000 for the normal and logistic cases, respectively. Cornfield and Mantel [1950] have shown that for the logistic tolerance distribution, the maximum likelihood estimator and the Spearman estimator were approximately equal and this algebraic approximation improved as $h \to 0$. Bross [1950] evaluates some sampling distributions via enumeration for the mle and the Spearman estimator (Se) using the logistic tolerance distribution and four dose levels, $n = 2$, and $n = 5$. He finds that the Se is more closely concentrated around the true mean than the mle. Haley [1952] performs similar computational results for the normal distribution.

Problems

(1) Show that the asymptotic efficiency of the Spearman estimator is equal to
(a) 0.9814 when the tolerance distribution is normal,

(b) 0.8106 when the tolerance distribution is angular: its d.f. is

$$F(x;\mu) = \sin^2(x - \mu + \pi/4), \quad -\frac{\pi}{4} \leqslant x - \mu \leqslant \frac{\pi}{4};$$

(c) 0 when the d.f. is one particle, i.e.

$$F(x,\mu) = 1 - \exp(-x/\mu), \, x > 0;$$

(d) $0.5 \leqslant E \leqslant 0.8319$ when the d.f. is algebraic, i.e.

$$F(x;\mu) = 1 - x^{-s}, \, x, s \geqslant 1;$$

(e) not defined when the d.f. is uniform, i.e.

$$F(x,\mu) = \tfrac{1}{2} + x - \mu, \, -\tfrac{1}{2} \leqslant x - \mu \leqslant \tfrac{1}{2};$$

(Hint: $F(x,\mu)$ is not differentiable for all values of μ.)

(f) 1.00 when the d.f. is logistic, i.e.

$$F(x;\mu) = [1 + e^{-(x-\mu)}]^{-1}, \, -\infty < x < \infty.$$

Hint for (a): After partial integration once we get

$$\int_{-\infty}^{\infty} \Phi(1-\Phi)dx = 2\int_{-\infty}^{\infty} x\Phi\phi dx = -2\Phi\phi\Big|_{-\infty}^{\infty} + 2\int_{-\infty}^{\infty}\phi^2 dx = 1/\sqrt{\pi}.$$

Also numerical integration yields

$$\int_{-\infty}^{\infty} \phi^2[1-\Phi]^{-1}dx = 0.903.$$

Hint for (b): $\int_0^{\pi/2} \sin^2 u \cos^2 u \, du = \pi/16.$

(2) Let

$$P_e(x) = K(e) \int_{-\infty}^{x} \left[1 + \frac{(t-\mu)^2}{1+2e}\right]^{-1-e} dx, \quad \text{where } K(e) = \Gamma(1+e)/\sqrt{\pi(1+2e)}\,\Gamma(1+e/2).$$

Show that the asymptotic efficiency of the Spearman estimator for the above tolerance distribution is arbitrarily close to zero and will tend to zero as $e \to 0$.
Hint: Show that

$$\int_{-\infty}^{\infty} P_e(x)[1 - P_e(x)]dx \geqslant K(e)2^{-1-e}[2e(1+2e)]^{-1} \quad \text{and}$$

$$\int_{-\infty}^{\infty} \left[\frac{\partial P_e(x)}{\partial \mu}\right]^2 \{P_e(x)[1 - P_e(x)]\}^{-1} dx \geqslant 8K^2(e)2^{-2-2e}(3+4e)^{-1}].$$

4.11.4. *Derivation of Fisher's Information for the Spearman Estimator*

Let the dose levels be

$$x_i = x_0 + ih \quad (i = 0, \pm 1, \ldots, \pm k),$$

where x_0 is chosen at random in an interval $(0, h)$ for a specified dose interval h. Let n subjects be administered dose levels x_i $(i = 0, \pm 1, \ldots, \pm k)$. If $F(x - \mu)$ denotes the tolerance distribution, the Fisher's information for this experiment is

$$I_k(x_0) = E[(\partial \ln L/\partial \mu)^2 | x_0]$$

where

$$L = \prod_{i=-k}^{k} \binom{n}{r_i} [F(x_i - \mu)]^{r_i} [1 - F(x_i - \mu)]^{n-r_i}.$$

Then

$$\frac{\partial \ln L}{\partial \mu} = \sum_{i=-k}^{k} \frac{(\partial F_i/\partial \mu)[r_i - nF(x_i - \mu)]}{F(x_i - \mu)[1 - F(x_i - \mu)]}.$$

Since the r_i are independent, it easily follows that

$$I_k(x_0) = \sum_{i=-k}^{k} \frac{n[\partial F(x_i - \mu)/\partial \mu]^2}{F(x_i - \mu)[1 - F(x_i - \mu)]}.$$

For an infinite experiment, that is, when $k \to \infty$, the information is

$$I = \lim_{k \to \infty} E_{x_0}[I_k(x_0)]$$

$$= \lim_{k \to \infty} \frac{1}{h} \int_0^h \sum_{-k}^{k} \frac{n[\partial F(x_i - \mu)/\partial \mu]^2}{F(x_i - \mu)[1 - F(x_i - \mu)]} dx_0$$

$$= \frac{n}{h} \lim_{k \to \infty} \sum \int_{ih}^{(i+1)h} \frac{[\partial F(x - \mu)/\partial \mu]^2}{F(x - \mu)[1 - F(x - \mu)]} dx$$

$$= \frac{n}{h} \int_{-\infty}^{\infty} \frac{f^2(x)}{F(x)[1 - F(x)]} dx = \frac{n}{h} E\left\{ \frac{f(X)}{F(X)[1 - F(X)]} \right\}.$$

Notice that I is free of the location parameter μ.

4.12. Reed-Muench Estimate

We assume the dose levels are equally spaced with spacing h. n denotes the common number of experimental units at each dose level. Also, let r_i denote the number of responses at the i-th dose level $(i = 1, \ldots, k)$. Let

$$s = \max\left[j: 1 \leq j \leq k \text{ and } \sum_1^j r_i \leq \sum_{j+1}^k (n - r_i) \right].$$

Define

$$\hat{\mu}_{RM} = x_k - h(k - s).$$

It has been pointed out (see for instance Miller [1973]) that by setting $\hat{p}_k = \frac{1}{2}$ in the Spearman-Karber estimator, one gets the Reed-Muench estimator.

4.13. *Dragstadt-Behrens Estimator*

With the above notation let

$$\hat{p}_s = \sum_1^s r_i \bigg/ \left[\sum_1^s r_i + \sum_{i=s+1}^k (n - r_i) \right] \quad (s = 1, \ldots, k).$$

If there exists an s such that $\hat{p}_s = \frac{1}{2}$, then define $\hat{\mu}_{DB} = x_k - h(k - s)$. If there is more than one s, take the average of the s values.

Remark. It seems that the Reed-Muench and Dragstadt-Behrens estimators are equivalent. However, they are inefficient in comparison with the Spearman-Karber estimator. Also notice that

$\hat{p}_s = \frac{1}{2}$ implies that $\sum_1^s r_i = \sum_{i=s+1}^k (n - r_i)$, which in turn implies that

$$k - s = \sum_1^k (r_i/n) = \sum_1^k \hat{p}_i.$$

4.14. *Application of the Spearman Technique to the Estimation of the Density of Organisms*

Johnson and Brown [1961] have given a procedure for estimating the density of organisms in a suspension by a serial dilution assay. The estimator based on the Spearman technique is simple and easy to compute.

To estimate the density of a specific organism in a suspension, a common method is to form a series of dilutions of the original suspension. Then from each dilution, a specified volume (hereafter be called a dose) is placed in each of several tubes. The tubes are later examined for evidence of growth of the organism.

Let $Z_i = a^{-i} Z_0$ denote the i-th dilution of Z_0, the highest concentration of original suspension and $a > 1$ denote the dilution factor $(i = 0, \ldots, k)$; let $x_i = \ln Z_i$ $(i = 0, \ldots, k)$. Suppose n tubes receive a unit volume (dose) for each dilution. Let r_i denote the number of tubes showing signs of growth at dilution Z_i, $s_i = n - r_i$, $p_i = r_i/n$ and $q_i = 1 - p_i$ $(i = 0, \ldots, k)$. The probability of signs of growth (i.e. the probability of one or more organism) is

$$P(Z_i) = 1 - \exp(-\theta Z_i) = 1 - \exp(-\theta e^{x_i}) = F(x_i) \text{(say)}.$$

The maximum likelihood method for estimating θ will be quite tedious. Suppose we are interested in estimating the mean μ of $F(x)$. The Spearman-Karber estimator for μ is given by

$$\hat{\mu} = x_0 + (d/2) - d \sum_{i=0}^k p_i$$

where $x_0 = \ln Z_0$, $d = \ln a$, and p_i is the sample proportion of tubes having growth at dilution Z_i.

Also,

$$\mu = \int_{-\infty}^{\infty} x\theta e^x \exp(-\theta e^x) dx = \int_0^{\infty} (\ln y - \ln \theta) e^{-y} dy = -\gamma - \ln \theta,$$

where $\gamma = 0.57722$ is Euler's constant. Then, we can express the parameter of interest θ as

$$\theta = \exp(-\gamma - \mu)$$

and we propose to use as point estimator

$$\hat{\theta} = \exp(-\gamma - \hat{\mu}) = \exp(-\gamma - x_0 - d/2 + d\sum p_i).$$

That is,

$$Z_0 \hat{\theta} = \exp(-\gamma - d/2 + d\sum p_i).$$

Furthermore,

$$\operatorname{var}(\hat{\mu}|x_0) = d^2 n^{-1} \sum F(x_i)[1 - F(x_i)]$$

$$\simeq \frac{d}{n} \int_{-\infty}^{\infty} F(x)[1 - F(x)]dx = -\frac{d}{n}\int_0^1 \frac{(1-u)}{\ln u} du = \frac{d \ln 2}{n},$$

since

$$\int_0^1 [x^b - x^a]/\ln x \, dx = \ln[(1+b)/(1+a)] \quad \text{for a, b} > -1.$$

$$E(\hat{\mu}|x_0) = x_0 + (d/2) - d\sum F(x_i)$$
$$= x_0 + (d/2) - d\sum [1 - \exp(-\theta e^{x_0 - id})].$$

Hence

$$B(\hat{\mu}|x_0) = E(\hat{\mu}|x_0) - \mu = x_0 + (d/2) + \gamma + \ln\theta - d\sum[1 - \exp(-e^{x_0 - id})].$$

Numerical computations carried out by the authors indicate that if the range of doses is wide enough and the dilution factor a is 10 or less, then bias is approximately zero. Since $\hat{\mu}$ is asymptotically normal for large n (the common number of tubes at each dose level), the 95% confidence interval for μ is

$$\hat{\mu} \pm 1.96(d\ln 2/n)^{1/2}$$

and the corresponding interval estimator for θ is

$$\hat{\theta}\exp[-1.96(d\ln 2/n)^{1/2}] < \theta < \hat{\theta}\exp[1.96(d\ln 2/n)^{1/2}].$$

Johnson and Brown [1961] consider a uniform distribution for x_0 on $(A, A + d)$ and find that if the doses extend over a wide enough range

$$E(\hat{\mu}) \doteq \mu \quad \text{and} \quad \operatorname{var}(\hat{\mu}) \doteq d\ln 2/n.$$

Also,

$$E(\hat{\theta}) = E(e^{-\gamma-\hat{\mu}}) = E[e^{-(\gamma+\mu)-(\hat{\mu}-\mu)}] = \theta \; E[e^{-(\hat{\mu}-\mu)}] = \theta \; E\left[\sum_{j=0}^{\infty}(-1)^j(\hat{\mu}-\mu)^j/j!\right].$$

Ignoring central moments 3 and higher of $\hat{\mu}$, we have

$$E(\hat{\theta}) \doteq \theta\left[1 + \frac{d\ln 2}{2n}\right].$$

Thus, a less biased estimate of θ would be

$$\hat{\theta}' = 2n(2n + d\ln 2)^{-1}\hat{\theta}.$$

The coefficient of variation of $\hat{\theta}$, CV can be approximated as

$$CV(\hat{\theta}) = \sqrt{var(\hat{\theta})}/\theta \doteq [\theta^2 \, var(\hat{\mu})]^{1/2}/\theta \doteq [d\ln 2/n]^{1/2}.$$

The authors note that the $var(\hat{\mu}) \doteq d\ln 2/n$ coincides with the asymptotic variance of the estimator for $\ln\theta$ proposed by Fisher [1922]. Thus, the asymptotic efficiency of the two procedures for estimating θ will be the same, namely 88%. In choosing the range of dilutions, the conditions $F(x_0) \geq 0.99$ and $F(x_0 - kd) \leq 0.01$ should be satisfied. That is, the expected number of organisms per dose at the highest and lowest concentration should be greater than 5 and less than 0.01, respectively.

Example [Johnson and Brown, 1961, table 1]

Consider the following data. Suppose it is known that the density of organisms of θ is between 10^2 and 10^5 organisms per unit volume. Let a 10-fold dilution factor be used. In order to ensure spanning the range from 0.01 to 5 organisms per dose, it is necessary to use 7 dilutions $(10^{-1},\ldots,10^{-7})$ of the original suspension.

Dilution	10^{-1}	10^{-2}	10^{-3}	10^{-4}	10^{-5}	10^{-6}	10^{-7}
Proportion of tubes showing growth	3/3	3/3	2/3	0/3	0/3	0/3	0/3

Thus

$n = 3, \sum p_i = 8/3, d = \ln 10.$

$$\hat{\mu} = \ln 10^{-1} + \frac{\ln 10}{2} - (\ln 10)(8/3) = -7.2915.$$

Hence $\hat{\theta} = e^{-\gamma-\hat{\mu}} = 824$ organisms per unit volume of the original suspension. Since $\exp[1.96(d\ln 2/n)^{1/2}] = 4.2$, the 95% confidence interval is $196 = 824/4.2 < \theta < (824)(4.2) = 3,460$. Also $\hat{\theta}' = 2n(2n + d\ln 2)^{-1}\hat{\theta} = (0.79)(824) = 651$ organisms per unit volume.

4.15. Quantit Analysis (Refinement of the Quantal Assay)

Earlier we dealt with parametric methods for analysing quantal response assays based on normal, logistic, uniform and other two-parameter families of tolerance distribution. Besides LD_{50}, extreme percentages are of interest. For example in sterilization tests for fruit flies, the quarantine officials require small ($<10\%$) survival rates. In therapeutics one is interested in determination of the safety margin, namely the difference between curative and lethal doses. Here we wish to estimate ED_{99} (the dose that cures 99%) and/or the LD_{01} (the dose that kills 1%). However, one requires a large sample size to estimate the extreme doses or we need a more precise mathematical model for the dose-response function. In general, asymmetric tolerance distribution will be adequate as a dose-response function provided logarithmic transformation of the dosage is used. Copenhaver and Mielke [1977] propose the omega distribution as a model for the tolerance distribution given by

$$F[x(\theta)] = \theta, \quad f[x(\theta)] = 1 - |2\theta - 1|^{\nu+1} \quad (0 < \theta < 1, \nu > -1).$$

Since $x(\theta) = F^{-1}(\theta)$, $dx = [f(F^{-1}(\theta))]^{-1} d\theta = [f(x(\theta))]^{-1} d\theta$,

$$x(\theta) = \int_{1/2}^{\theta} [f(x(z))]^{-1} dz.$$

Special cases: $\nu = 1$, gives $f(x(\theta)) = 4\theta(1 - \theta)$.

$x(\theta) = (1/4)\ln[\theta/(1 - \theta)]$, which is the logit of θ.

$F(x) = (1 + e^{-4x})^{-1}$.

$\nu = 0$ gives the double exponential density given by $f(x) = \exp(-2|x|)$. As $\nu \to \infty$, $f(x) = 1$, $-1/2 < x < 1/2$.

When ν is near 2, the shape of this distribution is similar to that of the normal distribution. Prentice [1976] considered a family of tolerance distributions given by

$$f(x) = e^{x\gamma_1}(1 + e^x)^{-(\gamma_1+\gamma_2)}/B(\gamma_1, \gamma_2) \quad (\gamma_1\gamma_2 > 0) \quad \text{and}$$

$B(a,b) = \Gamma(a)\Gamma(b)/\Gamma(a + b)$.

If $\gamma_1 = \gamma_2 = \gamma$ and $\gamma \to 0$, we get the double-exponential density and if $\gamma \to \infty$, we get the normal density, if $\gamma = 1$, we get the logistic density. The omega includes all the symmetric distributions included in the class of Prentice [1976] if $0 < \nu < 2$. The latter, of course, includes many asymmetrical distributions when $\gamma_1 \ne \gamma_2$.

If P_i is the probability of responding to the dose level x_i, then $P_i = F(\alpha + \beta x_i)$ and the tolerance distribution is given by

$$f(\alpha + \beta x_i) = 1 - |2P_i - 1|^{\nu+1}.$$

Furthermore

$$\alpha + \beta x_i = h_v(P_i) = \int_{1/2}^{P_i} (1 - |2z - 1|^{v+1})^{-1} dz.$$

where $h_v(P_i)$ is called the 'quantit' of P_i and $h_v(p_i)$ is the observed quantit corresponding to p_i which equals the sample proportion responding to dose level x_i. Some adjustment is made to p_i when all or no subjects respond.

4.15.1. Computational Procedure for Estimating the Parameters

First an initial value $v_0 = 1$ is used for v and α and β are, as the least squares, solutions of $h_1(p_i) = \alpha + \beta x_i$ ($i = 1, \ldots, k$). An efficient search routine is used for determining $\hat{v}$. The iterative process is continued until an accuracy of two decimal places for v is obtained. If $\hat{v}$ exceeds 20, $\hat{v}$ is set at 20 [if $\hat{v} \geqslant 20$, the omega distribution resembles the uniform on $(-\frac{1}{2}, \frac{1}{2})$]. The iterative method of maximum likelihood is simplified if the second-order partial derivatives of the likelihood function are replaced by their estimated expectations (this is called method of scoring). The quantit $h_v(P)$ can be expressed as infinite series given by

$$h_v(P) = \int_{1/2}^{P} [1 - |2z - 1|^{v+1}]^{-1} dz = (\delta/2) \sum_{j=0}^{\infty} |2P - 1|^{j(v+1)+1} \Big/ [j(v + 1) + 1], \quad \text{where}$$

$$\delta = \begin{cases} +1 & \text{if } P > \frac{1}{2} \\ 0 & \text{if } P = \frac{1}{2} \\ -1 & \text{if } P < \frac{1}{2}. \end{cases}$$

(Note that the above series is obtained by expanding in power series of $|2z - 1|^{v+1}$ and integrating term by term.) Hence numerical integration can be avoided in computing $h_v(P)$. The value of $h_v(P)$ can be obtained by summing the first M terms of the series and then adding a suitable remainder term R_M, where

$$R_M = \int_{M+1/2}^{\infty} \frac{|2P - 1|^{w(v+1)+1}}{w(v + 1) + 1} dw = (v + 1)^{-1} \int_{\psi}^{\infty} (e^{-z}/z) dz$$

where $\psi = -[(M + \frac{1}{2})(v + 1) + 1] \ln|2P - 1|$ (Hint: set $[w(v + 1) + 1] \ln|2P - 1| = -z$.)

Copenhaver and Mielke [1977] recommend using a well-known continued fraction approximation for the last integral representation of R_M given by

$$\int_x^{\infty} (e^{-z}/z) dz = \exp(-x) \left[\frac{1}{x+} \frac{1}{1+} \frac{1}{x+} \frac{2}{1+} \frac{2}{x+} \frac{3}{1+} \frac{3}{x+} \cdots \right] \quad (x > 0).$$

Magnus et al. [1977] provide a faster convergent series for $h_v(P)$ than the above approximation.

A measure of the goodness-of-fit of the maximum likelihood solution is the value of the likelihood which also provides a test-of-fit for the logistic model. Let $L(\omega)$

and $L(\Omega)$ denote the value of the likelihoods for logit and quantit analyses, respectively, where the likelihood L is given by

$$L = \prod_{i=1}^{k} \binom{n_i}{r_i} P_i^{r_i}(1-P_i)^{n_i-r_i}, \quad \text{where } P_i = F(\alpha + \beta x_i).$$

Then for testing $H_0: v = 1$, we use

$$T = 2[\ln L(\Omega) - \ln L(\omega)]$$

which is asymptotically distributed as χ^2 with 1 d.f.

Copenhaver and Mielke [1977] analysed 22 sets of published data with the quantit method. Except for a couple of data sets, the omega model yields smaller χ^2 goodness-of-fit (given by $\chi^2 = \sum_{i=1}^{k} n_i(p_i - P_i)^2 / P_i(1-P_i)$ $(p_i = r_i/n_i, i = 1,\ldots,k)$ and smaller $L(\Omega)$. The authors also tabulate the estimates and standard errors of the $\log ED_{50}$ and ED_{99} for each of the logit, probit and quantit models. Although ED_{50} estimates are in close agreement, significant differences exist among the ED_{99} estimates, especially for those data sets in which the ML estimates of v is either near -1 or much larger than 1. Note that the omega distribution has lighter (heavier) tails than the logistic distribution when v is greater (less) than 1. Hence, for those omega model solutions with $\hat{v}$ much greater (less) than 1, the logistic model will very likely over(under)estimate extreme dose levels such as ED_{99} and its standard error. In other words, in estimating the lower or upper extreme percentiles, the omega model estimates will be closer to (farther from) the $\log ED_{50}$ estimate than the logistic model estimates when $\hat{v}$ is greater (less) than 1.

4.15.2. *Polychotomous Quantal Response*

The quantal response in a biological assay is usually based on two possible outcomes. A common example of the two possible outcomes is death or survival. However, there are several assays, such as that of some insecticide based on the mortality of the housefly. There are three possible outcomes, namely, active, moribund, and dead. There is a definite order of these outcomes: active precedes moribund, which in turn precedes dead. It may be more efficient to analyze the data with more than two outcomes, than combining the outcomes into two categories. An approach to the problem based on the method of maximum likelihood has been given by Aitchison and Silvey [1957] and Ashford [1959]. Gurland et al. [1960] provide a general method of analyzing the data when one uses the normits (probits), logits or other monotone transformation. The estimates of the parameters used in the general method are obtained by minimizing the appropriate χ^2 expressions and have the asymptotic properties of consistency and efficiency. The latter set of authors provide a numerical example involving the housefly insecticides.

4.16. *Planning a Quantal Assay*

If μ denotes the mean and σ^2 the variance of the tolerance distribution, Brown [1970] discusses the design aspect for estimating LD_{50} of a single preparation within R-fold of its true value. That is, $\hat{\mu}$ should be between μ/R and μR. Assume that bounds μ_1 and μ_2 and σ_1 and σ_2 exist such that

$$\mu_1 \leqslant \log LD_{50} \leqslant \mu_2 \quad \text{and} \quad \sigma_1 \leqslant \sigma \leqslant \sigma_2.$$

Note that the standard deviation can be interpreted as roughly one third to one sixth of the 'range' of the dose-response function. Since it is prudent to use a range of doses that ensures a low dose with a low probability of response and a high dose with a high probability of response, Brown [1966, 1970] recommends choosing

$$x_1 = \mu_1 - 2\sigma_2, \quad x_k = \mu_2 + 2\sigma_2.$$

In order to ensure negligible bias due to a coarse dilution factor, the doses should be spaced no more than 2σ. Thus, h, the distance on the log scale, should be $h = 2\sigma_1$.

The variance of the Spearman-Karber estimator is

$$\frac{h^2}{n} \sum_{-\infty}^{\infty} F\left(\frac{x_i - \mu}{\sigma}\right)\left[1 - F\left(\frac{x_i - \mu}{\sigma}\right)\right] = \frac{h}{n\sigma} \int_{-\infty}^{\infty} F(x)[1 - F(x)]dx$$

$$= h/n\sigma \quad \text{if logistic}$$

$$= h/n\sigma\sqrt{\pi} \quad \text{if normal}.$$

Hence, one can approximately take the standard deviation of the Spearman-Karber estimate to be

$$SD(\log LD_{50}) = (h\sigma/2n)^{1/2}.$$

Setting two standard deviations equal to the desired error limit, $\log R$ gives

$$(\log R) = 2(h\sigma/2n)^{1/2}.$$

Setting $h = 2\sigma_1$ and since $\sigma \leqslant \sigma_2$

$$n \leqslant 4\sigma_1\sigma_2/(\log R)^2.$$

For example, suppose an assay of the ED_{50} of an estrogen preparation be precise within 1.6 times the true value with 95% confidence. Also assume that

$$-1.00 \leqslant \mu \leqslant 0.0 \quad \text{and} \quad 0.40 \leqslant \sigma \leqslant 0.60.$$

Hence $x_1 = -1.00 - 1.20 = -2.20$, $x_k = 0.0 + 1.20 = 1.20$, $h = 2(0.4) = 0.8$. Thus, the log doses are

$x_i = 1.20, 0.40, -0.40, -1.20, -2.00, -2.80$

and the number of subjects at each dose level should be

$n = 4(0.4)(0.6)/(0.2)^2 = 24$.

Maximal Information from Bioassays

Emmens [1966] draws the following conclusions based on empirical studies. It is advisable to utilize graded responses such as body weight, organ weight, percentage of blood sugar, or any measure that can be quantitatively expressed, rather than quantal responses. Categorizing responses into grades 1, 2, 3, and 4 will usually be superior to a purely quantal measurement. The weights used in the probit analysis of quantal responses illustrate that a graded response provides, on the average, about twice the information of a quantal response. Also, graded response is more useful because the latter can be measured over a wider range than quantal responses without running into extremes.

Quantal responses based on groups of animals are preferable to a series of single observations at several dose levels. Also, observations derived serially from the same animal may be four to six times as precise as observations made simultaneously upon different animals. If it is impossible to take replicated observation on the same animal, the use of littermates or sometimes of genetically homogeneous material is helpful.

In large scale assays, especially those which extend over a period of time, some animals may die or some observations may be lost. Emmens [1966] recommends including one or two extra animals per group and to reject results at random from those groups whose eventual number is greater than the smallest. This method will yield an unbiased and balanced experiment as long as animals do not die because of treatment effects.

4.17. *Dose Allocation Schemes in Logit Analysis*

A common experimental design is to select a number of dose levels and administer each dose level to an equal number of subjects. Optimal designs have been discussed in the literature. Tsutakawa [1980] employed Bayesian methods to select dose levels for the logistic response function. Hoel and Jennrich [1979] showed that for a k-parameter response function, the optimal design has k dose levels. Hence, McLeish and Tosh [1985] consider two dose levels for the logit response function and search for optimal dose levels when cost constraints are placed on the experiment. The total cost is equal to the number of observations plus a constant times the expected number of 'deaths' or responses. Thus, experimental designs having the same variance but which reduce the expected number of responses have cost advantages over classical designs.

In the logit model, a subject is administered a dose x and the response (Y = 1) or no response (Y = 0) of the subject is observed. The probability of response is

$$P(Y = 1|x) = 1 - P(Y = 0|x) = P(x), \quad \text{where}$$

$$P(x) = (1 + e^{-\alpha - \beta x})^{-1}. \tag{1}$$

Alternative parametrization of P is

$$P(x) = \left[1 + \frac{q_r}{p_r} e^{-\beta(x-r)}\right]^{-1} \tag{2}$$

where β is an unknown scale parameter, r is an unknown root of the equation $P(x) = p_r$ and $p_r = 1 - q_r$ is a known constant ($0 < p_r < 1$). We can also parametrize p in terms of unknown fractiles p_1 and p_2 and known doses x_1 and x_2 which satisfy

$$p_i = \left[1 + \frac{q_r}{p_r} e^{-\beta(x_i - r)}\right]^{-1} \quad (i = 1, 2) \tag{3}$$

The usual cost restraint for a two-dose experiment with n_i observations at x_i (i = 1, 2) is

$$c = n_1 + n_2. \tag{4}$$

A more general constraint is

$$c = n_1 + n_2 + D(n_1 p_1 + n_2 p_2), \tag{5}$$

where D is the 'extra' cost of a response. For instance, when response corresponds to the death of an experimental unit due to the toxicity of the drug administered, the cost of the experiment would be the number of observations plus a constant times the number of deaths occurred. McLeish and Tosh [1985] consider the constraint (5); however, they obtain explicit solutions for (4) only. For the sake of simplicity we shall confine to constraint (4).

Optimal Sampling Strategy

We wish to estimate the root r of the equation $P(x) = p_n$ for given p_n. A design is said to be optimal if it minimizes the variance of r subject to the restriction (4). If we parametrize p in terms of (3), the information matrix becomes

$$I\begin{vmatrix} p_1 \\ p_2 \end{vmatrix} = \begin{vmatrix} n_1(p_1 q_1)^{-1} & 0 \\ 0 & n_2(p_2 q_2)^{-1} \end{vmatrix}.$$

Dose Allocation Schemes in Logit Analysis

Reparametrization in terms of (2) gives the information matrix

$$I\begin{vmatrix}r\\ \beta\end{vmatrix} = \begin{vmatrix}\beta^2(w_1 + w_2) & \beta(x_1 - r)w_1 + \beta(x_2 - r)w_2\\ \beta(x_1 - r)w_1 + \beta(x_2 - r)w_2 & (x_1 - r)^2 w_1 + (x_2 - r)^2 w_2\end{vmatrix} \quad (6)$$

where $w_i = n_i p_i q_i$ ($i = 1, 2$).

The 1-1 element of the inverse of $I\begin{vmatrix}r\\ \beta\end{vmatrix}$ is

$$V = \beta^{-2}(x_2 - x_1)^{-2}\left[\frac{(x_1 - r)^2}{n_2 p_2 q_2} + \frac{(x_2 - r)^2}{n_1 p_1 q_1}\right]. \quad (7)$$

Note that V is the asymptotic variance of the maximum likelihood estimate $\hat{r}$ of r. Thus, we wish to minimize V subject to (4). So, define

$$Q = V + \lambda(n_1 + n_2 - c)$$

$$\frac{\partial Q}{\partial n_1} = -\beta^{-2}(x_2 - x_1)^{-2}\frac{(x_2 - r)^2}{n_1^2 p_1 q_1} + \lambda = 0$$

$$\frac{\partial Q}{\partial n_2} = -\beta^{-2}(x_2 - x_1)^{-2}\frac{(x_1 - r_1)^2}{n_2^2 p_2 q_2} + \lambda = 0.$$

Notice that

$$\lambda n_1 + \lambda n_2 = V, \quad \text{so that } \lambda = (V/c).$$

Hence

$$n_1 = \left(\frac{c}{V p_1 q_1}\right)^{1/2} \frac{|x_2 - r|}{\beta|x_2 - x_1|} \quad \text{and}$$

$$n_2 = \left(\frac{c}{V p_2 q_2}\right)^{1/2} \frac{|x_1 - r|}{\beta|x_2 - x_1|}. \quad (8)$$

Substitution of n_1 and n_2 in (7) gives

$$V = c^{-1}\beta^{-2}(x_2 - x_1)^{-2}\left[\frac{|x_2 - r|}{(p_1 q_1)^{1/2}} + \frac{|x_1 - r|}{(p_2 q_2)^{1/2}}\right]^2. \quad (9)$$

Now we determine the value of x_1 and x_2 which minimize (9). Towards this, let

$$h(x) = \{P(x)[1 - P(x)]\}^{-1/2}. \quad (10)$$

Then (9) becomes

$$V = c^{-1}\beta^{-2}(x_2 - x_1)^{-2}[|x_2 - r|h(x_1) + |x_1 - r|h(x_2)]^2.$$

Define

$$\sigma = V^{1/2} = [|x_2 - r|h(x_1) + |x_1 - r|h(x_2)]/c^{1/2}\beta|x_2 - x_1|. \quad (11)$$

Without loss of generality assume that $x_1 < x_2$. Then we have the alternate forms for σ.

$$\sigma = \frac{(x_2 - r)}{\beta\sqrt{c}}\left[\frac{h(x_2)\operatorname{sgn}(x_1 - r) + h(x_1)\operatorname{sgn}(x_2 - r)}{x_2 - x_1}\right] - \frac{h(x_2)\operatorname{sgn}(x_1 - r)}{\beta\sqrt{c}}$$

$$= \frac{(x_1 - r)}{\beta\sqrt{c}}\left[\frac{h(x_2)\operatorname{sgn}(x_1 - r) + h(x_1)\operatorname{sgn}(x_2 - r)}{x_2 - x_1}\right] + \frac{h(x_1)\operatorname{sgn}(x_2 - r)}{\beta\sqrt{c}}. \quad (12)$$

The partial derivative of the first form w.r.t. x_1 gives

$$h(x_1) + [\operatorname{sgn}(x_1 - r)(x_2 - r)]h(x_2) = -(x_2 - x_1)h'(x_1). \quad (13)$$

The partial derivative of the second form w.r.t. x_2 gives

$$h(x_1) + [\operatorname{sgn}(x_1 - r)(x_2 - r)]h(x_2) = (x_2 - x_1)h'(x_2). \quad (14)$$

Now using the fact that $dP(x)/dx|_{x=x_i} = p_i q_i \beta$, we find that

$$h'(x_i) = (p_i - q_i)\beta/2(p_i q_i)^{1/2} \quad (i = 1, 2). \quad (15)$$

Using (15) in (13) and (14) we obtain

$$(p_1 - q_1)/(p_1 q_1)^{1/2} = -(p_2 - q_2)/(p_2 q_2)^{1/2}. \quad (16)$$

Now let

$$g(p) = (2p - 1)(p - p^2)^{-1/2}.$$

Since

$$g'(p) = [4 + (2p - 1)^2]/(p - p^2) > 0 \quad \text{for all } p,$$

$g(p)$ is monotone increasing and is symmetric about $p = \frac{1}{2}$. Hence, the only solutions of (16) are of the form $p_1 = q_2$ or equivalently $x_1 = -x$, $x_2 = x$ [since $x_1 = \ln(p_1/q_1)$ and $x_2 = \ln(p_2/q_2)$]. Also, using the fact that $r < x_1 < x_2$, when $\alpha = 0$ and $\beta = 1$, the equation becomes $h(-x) + h(x) = 2xh'(x)$. Since $h(-x) = h(x)$, we get $x/2 = (p_2 - q_2)^{-1}$, i.e., $x/2 = (1 + e^{-x})/(1 - e^{-x})$. Then the specific solutions are

$x_1 = -2.39935$ and $x_2 = 2.39935$ for which

$p_1 = 0.08322$ and $p_2 = 0.91678$.

Then the optimal strategy is to sample at the root r if $p_1 \leq p_r \leq p_2$ and sample at $P^{-1}(p_1)$ and $P^{-1}(p_2)$ when either $p_r < p_1$ or $p_2 < p_r$.

Remarks. (1) If we consider constraint (5), then $h(x) = \{[1 + DP(x)]/P(x)[1 - P(x)]\}^{1/2}$. McLeish and Tosh [1985], using the convexity of $h(x)$, show the existence of optimal dose levels. (2) Generally, one hopes that the optimal asymptotic design would be practical for small sample sizes. Here it is not the case, because sampling entirely at the root or at widely spaced quantiles is inefficient for smaller sample sizes.

Appendix: Justification for Anscombe's Correction for the Logits l_i[2]

Let n_i be the number of experimental units getting the dose x_i and let R_i denote the number responding to that dose level ($i = 1, \ldots, k$). Then let

$$Z_i(\delta) = \ln[(R_i + \delta)/(n_i - R_i + \delta)]$$

we shall find δ such that $Z_i(\delta)$ are asymptotically unbiased for $\alpha + \beta x_i$. Hereafter let us drop the subscript i throughout. Let

$$R = nP + Un^{1/2}.$$

Then $EU = 0$ and $\text{var} U = P(1 - P)$ and U is bounded in probability. Let $\lambda = \ln[P/(1 - P)]$. Consider

$$Z(\delta) - \lambda = \ln(R + \delta) - \ln nP - \{\ln(n - R + \delta) - \ln[n(1 - P)]\}$$

$$= \ln\left(1 + \frac{U}{Pn^{1/2}} + \frac{\delta}{nP}\right) - \ln\left[1 - \frac{U}{(1 - P)n^{1/2}} + \frac{\delta}{(1 - P)n}\right]$$

$$= \frac{U}{P(1 - P)n^{1/2}} + \frac{\delta(1 - 2P)}{P(1 - P)n} - \frac{(1 - 2P)U^2}{2P^2(1 - P)^2 n} + o_p(n^{-1}),$$

where $o_p(n^{-1})$ denote terms of smaller order of magnitude than n^{-1} in probability. Thus

$$EZ(\delta) - \lambda = (1 - 2P)\left(\delta - \frac{1}{2}\right)[P(1 - P)n]^{-1} + o(n^{-1}).$$

Hence $Z(\delta)$ will be almost unbiased for λ when n is sufficiently large, provided $\delta = 1/2$. Thus, a single choice of δ which is free of P is $1/2$. We set

$$Z = \log\left[\left(R + \frac{1}{2}\right)\bigg/\left(n - R + \frac{1}{2}\right)\right].$$

Gart and Zweifel [1967] have, by using similar methods, proposed $V = (n + 1)(n + 2)/n(R + 1)(n - R + 1)$ such that $\text{var} Z \doteq EV$, so that V is nearly unbiased for the variance of Z.

Problems

(1) The following data is taken from Morton [1942] which describes the application of retenone in a medium of 0.5% saponin, containing 5% of alcohol. Insects were examined and classified 1 day after spraying. Test of the toxicity of retenone to *Macrosiphoniella sanborni*:

---

[2] Cox [1970] served as a source for the material of the appendix.

The Logit Approach

Dose of rotenone mg/l	$\log_{10}$ dosage x	Number of insects, n	Number affected r	Proportion killed p
10.2	1.01	50	44	0.88
7.7	0.89	49	42	0.86
5.1	0.71	46	24	0.52
3.8	0.58	48	16	0.33
2.6	0.41	50	6	0.12
0		49	0	0

Because of the last row, we can assume that the natural mortality is zero. Estimate α and β by fitting the probit to the above data by the method of maximum likelihood. Starting with $b_0 = 4.01$ and $a_0 = -2.75$, obtain the maximum likelihood estimates of α and β.

(2) The following data is taken from Fisher and Yates [1963]:

x	3	4	5	6	7	8	9
n	8	8	8	8	8	8	8
Dead	0	0	2	3	3	7	8
p	0	0	0.25	0.375	0.375	0.875	1.000

Berkson [1957], fitting the logit by the method of maximum likelihood obtains (after the third iteration) $\hat{\alpha} = -8.0020$ and $\hat{\beta} = 1.2043$. He starts with the provisional values of $a_0 = -6.94$ and $b_0 = 1.04$.

Find a_0 and b_0 as the least-squares estimates based on $(x_i, \hat{l}_i)$, $i = 1, \ldots, k$ and then iterate by the method of maximum likelihood. Compare your values with those obtained by Berkson [1957]. Also, starting with Berkson's provisional values obtain the mle values of α and β.

(3) Let n subjects be given a dose x of a toxic substance and let r be the number responding to the same. Define

$l = \ln[(r + \frac{1}{2})/(n - r + \frac{1}{2})]$.

If $P(x) = [1 + e^{-(\alpha + \beta x)}]^{-1}$ where $P(x)$ is the probability of a single subject responding to dose level x, show that

$El = \alpha + \beta x + O(n^{-2})$

$Var l = (nPQ)^{-1} + O(n^{-2})$ where $Q = 1 - P(x)$.

Hint: Refer to Anscombe [1956].

5 Other Methods of Estimating the Parameters[1]

Wilson and Worcester [1943a, b, c] and Worcester and Wilson [1943] estimated the parameters of a logistic response curve when two or three dose levels with dilutions in geometric progression are used. Wilson and Worcester [1943d] considered the maximum likelihood estimation of the location and scale parameters of a general response curve.

The dose levels are said to be in geometric progression if they are of the form $D_1, D_2 = rD_1, D_3 = r^2 D_1, \ldots$ Taking their logarithms and setting $x_i = (\ln D_i - \ln D_1)/\log r$, we see that the x_i will take the values $0, 1, 2, \ldots$ Bacteriologists prefer working with dilutions to working with doses. Notice that dilutions and doses are inversely related: the lowest dilution corresponds to the largest dose level (and typically to the largest response).

5.1. Case of Two Dose Levels

Let D_1, $D_2 = rD_1$ be the two dose levels. Let the response at $x = \log D$ be denoted by P where

$$P = [1 + e^{-(\alpha + \beta x)}]^{-1} = [1 + e^{-\beta(x - \gamma)}]^{-1}, \quad \gamma = -\alpha/\beta.$$

Let $x_1 = \log D_1$, and $x_2 = x_1 + \log r$. If D_i denotes the response at x_i and $Q_i = 1 - P_i (i = 1, 2)$, then

$$\log(P_1/Q_1) = \beta(x_1 - \gamma) \quad \text{and} \quad \log(P_2/Q_2) = \beta(x_1 + \log r - \gamma).$$

Solving for β and γ (after setting $p_i = \hat{P}_i$ which is the sample proportion of n animals responding to dose level x_i (i = 1,2) we obtain

$$b = \hat{\beta} = [\log(p_2/q_2) - \log(p_1/q_1)]/\log r \quad \text{and}$$

$$c = \hat{\gamma} = [x_1 \log(p_2/q_2) - x_2 \log(p_1/q_1)]/[\log(p_2/q_2) - \log(p_1/q_1)].$$

If $l = \log(p/q)$, then $dl \doteq dp/PQ$ and consequently

$$\operatorname{var} b = [(P_1 Q_1)^{-1} + (P_2 Q_2)^{-1}]/n(\log r)^2.$$

---

[1] Ashton [1972] served as a source for part of this chapter.

Now, let $W_i = p_i q_i$, $S_i = \log(p_i/q_i)$, $i = 1, 2$.
Then applying the formula for the variance of a ratio, namely

$$\text{var}\left(\frac{U}{V}\right) = \theta_2^{-4}[\theta_2^2 \text{var } U + \theta_1^2 \text{var } V - 2\theta_1 \theta_2 \text{cov}(U,V)]$$

where $EU = \theta_1$ and $EV = \theta_2$, after setting

$$U = x_1 S_2 - x_2 S_1, \quad \text{and} \quad V = S_2 - S_1$$

we have

$$(S_2 - S_1)^4 s_c^2 = (S_2 - S_1)^2 [x_1^2 (np_2 q_2)^{-1} + x_2^2 (np_1 q_1)^{-1}]$$
$$+ (x_1 S_2 - x_2 S_1)^2 [(np_2 q_2)^{-1} + (np_1 q_1)^{-1}]$$
$$- 2(S_2 - S_1)(x_1 S_2 - x_2 S_1)[x_1 (np_2 q_2)^{-1} + x_2 (np_1 q_1)^{-1}].$$

That is,

$$n(S_2 - S_1)^4 s_c^2 = (S_2 - S_1)\left[(S_2 - S_1)\left(\frac{x_1^2}{W_2} + \frac{x_2^2}{W_1}\right) - (x_1 S_2 - x_2 S_1)\left(\frac{x_1}{W_2} + \frac{x_2}{W_1}\right)\right]$$
$$+ (x_1 S_2 - x_2 S_1)\left[(x_1 S_2 - x_2 S_1)(W_2^{-1} + W_1^{-1}) - (S_2 - S_1)\left(\frac{x_1}{W_2} + \frac{x_2}{W_1}\right)\right]$$

$$= (S_2 - S_1)\left[\frac{S_2 x_2^2}{W_1} - \frac{S_1 x_1^2}{W_2} - \frac{x_1 x_2 S_2}{W_1} + \frac{x_1 x_2 S_1}{W_2}\right]$$
$$+ (x_1 S_2 - x_2 S_1)\left[\frac{x_1 S_2}{W_1} - \frac{x_2 S_1}{W_2} - \frac{x_2 S_2}{W_1} + \frac{x_1 S_1}{W_2}\right]$$

$$= (S_2 - S_1)\left[\frac{S_2 x_2 (\ln r)}{W_1} + \frac{x_1 S_1 (\ln r)}{W_2}\right]$$
$$- (x_1 S_2 - x_2 S_1)\left[\frac{S_2 \ln r}{W_1} + \frac{S_1 \ln r}{W_2}\right], \quad \ln r = (x_2 - x_1).$$

Thus

$$\frac{n(S_2 - S_1)^4}{\ln r} s_c^2 = W_1^{-1}[S_2 x_2 (S_2 - S_1) - S_2 (x_1 S_2 - x_2 S_1)]$$
$$+ W_2^{-1}[S_1 x_1 (S_2 - S_1) - S_1 (x_1 S_2 - x_2 S_1)]$$
$$= W_1^{-1}[S_2^2 \ln r] + W_2^{-1}[S_1^2 \ln r].$$

Hence

$$s_c^2 = \frac{n^{-1}(\ln r)^2 \left\{\frac{[\ln(p_2/q_2)]^2}{p_1 q_1} + \frac{[\ln(p_1/q_1)]^2}{p_2 q_2}\right\}}{[\ln(p_1/q_1) - \ln(p_2/q_2)]^4}.$$

If there are two values of β, each determined by two doses on n animals, then

var(b − b′) = var b + var b′   and

var(c − c′) = var c + var c′.

If the values of β and β′ are different, the result is of lesser value than if β = β′, because standardization at the LD_{50} points does not guarantee standardization at other points.

When the response rates are 0 or 1, the corresponding logits l_i will become infinite in magnitude. A reasonable approach suggested by Berkson [1953, p. 584] is to replace p = 0 by $(2n)^{-1}$ and p = 1 by $1 - (2n)^{-1}$. This makes sense because the values of p increase in steps of $1/n$ and $(2n)^{-1}$ is halfway between 0 and n^{-1} and also $1 - (2n)^{-1}$ is halfway between 1 and $1 - n^{-1}$.

5.2. *Natural Mortality*

The question is how to incorporate natural mortality among experimental units. So far, we have assumed that the response has occurred solely because of the dose. However, this may not be true; for instance, in using insects, some may die even though not treated with the toxic. Also in tests on eggs, an adjustment has to be made for the proportion of eggs that fail to hatch due to natural causes.

In the case of an insecticide, let c denote the proportion of insects that would die due to natural causes, i.e., even in the absence of the toxic. Let P′ denote the proportion of deaths in the insects due to natural causes as well as the toxic. Let D denote the event of the death of an insect and T denote the effect of the toxic. Then

$$P = P(D|T) = P(DT)/P(T) = [P(D) - P(D\bar{T})]/P(T)$$
$$= (P' - c)/(1 - c).$$

If c is known or an estimate of it is available, one can make an adjustment with maximum likelihood or the logit using the above formula. So, replacing n_i by $n_i(1 - c)$ and p_i by $(p_i' - c)/(1 - c)$, the likelihood equations become

$\Sigma n_i(p_i' - P_i') = 0$,

$\Sigma n_i x_i (p_i' - P_i') = 0$.

Hence, the equations are unaltered. Notice that n_i and p_i' are the unadjusted observed values. Also since $\partial P'/\partial P = (1 - c)$

$\partial^2 L/\partial \alpha^2 = -(1 - c)\Sigma n_i W_i$, etc., where $W_i = P_i Q_i$,

the error variances of the estimates a and b are increased by the factor $(1 - c)^{-1}$ and consequently the precision is decreased. c = 0.2 implies that a natural mortality of 20% will increase the standard errors of a and b and ED_{50} by a factor of 1.1.

The above procedure is simple and approximately correct, although theoretically it is not very sound. The likelihood equations are obtained by replacing P by P' (Q by Q') in the expression for L since R follows the binomial distribution with parameter P'. Furthermore, it is P and not P' that follows the logistic law. Hence, the true likelihood equations are

$$\frac{\partial L}{\partial \alpha} = \frac{\Sigma n_i(p_i' - P_i')}{P_i' Q_i'} \frac{\partial P_i'}{\partial \alpha} = 0$$

$$\frac{\partial L}{\partial \beta} = \frac{\Sigma n_i(p_i' - P_i')}{P_i' Q_i'} \frac{\partial P_i'}{\partial \beta} = 0.$$

Recall that $\partial P/\partial \alpha = PQ$, $\partial P/\partial \beta = xPQ$ and note that

$$P' - p' = (1 - c)(p - P), \quad Q' = (1 - c)Q, \quad \partial P'/\partial P = 1 - c.$$

Using

$$\frac{\partial P'}{\partial \alpha} = \frac{\partial P'}{\partial P} \frac{\partial P}{\partial \alpha} = (1 - c)PQ,$$

we have for the true likelihood equations

$$\frac{\partial L}{\partial \alpha} = \Sigma n_i(p_i - P_i) \left[\frac{P_i}{P_i + c(1 - c)^{-1}} \right] = 0$$

$$\frac{\partial L}{\partial \beta} = \Sigma n_i x_i(p_i - P_i) \left[\frac{P_i}{P_i + c(1 - c)^{-1}} \right] = 0.$$

Since $P_i/[P_i + c(1 - c)^{-1}] < 1$, the weights attached to the observations are diminished. The simplicity of the earlier likelihood equations is lost, since

$$P_i(1 - c)/[P_i + c(1 - c)^{-1}] \doteq 1 - c - cP_i^{-1} \approx 1 - c \quad \text{when } P_i \text{ is large,}$$

Thus, the simple equations suggested earlier are reasonable unless quite a few P_i are small. Also differentiating the above likelihood equations and taking expectations we obtain

$$E \frac{\partial^2 L}{\partial \alpha^2} = -\Sigma n_i W_i g(P_i, c)$$

$$E \frac{\partial^2 L}{\partial \alpha \partial \beta} = -\Sigma n_i W_i x_i g(P_i, c)$$

$$E \frac{\partial^2 L}{\partial \beta^2} = -n_i W_i x_i^2 g(P_i, c), \quad \text{where}$$

$g(P_i, c) = P_i/[P_i + c(1 - c)^{-1}], \quad W_i = P_i Q_i.$

Hence, we infer that the precision of the estimates is diminished.

The likelihood equations are solved by using the iterative method suggested earlier. Now each of the quantities need to be multiplied by a factor $g(\hat{P}_i,c)$ where $c = r_c/n_c$ and is obtained from a controlled experiment in which r_c out of n_c insects die in the absence of the insecticide. Notice that we use the adjusted p_i in the place of the original observed p'_i. It is suggested that the above method is sufficiently accurate for $c \leqslant 0.10$. If $c > 0.10$ or the control batch is small then we should use an extended method of maximum likelihood procedure in which there are three unknown parameters. However, in practice, either we redo the control experiment so that the control batch is of a reasonable size or we cut short the duration of the experiment so that the resulting c is small.

Estimation of Parameters with Non-Zero Natural Mortality

Using GLIM one can iteratively fit the probit or logit model when the natural mortality is zero. Barlow and Feigl [1982, 1985] indicate how one can improvise the existing GLIM programs in order to fit the probit or the logit model with the non-zero natural rate c (which is known or estimated from a controlled experiment).

Example

In two alternative forced-choice psychophysical experiments of infant light perception, a light of known intensity is shown into the eyes of the infant from either the right or the left side; the side is randomly chosen. The response is correct if the baby's eyes shift to the other side. Since 'unknowns' are not permitted for the response, we use the terminology 'forced choice'. Different intensities of light will be termed different 'dose' levels. It is expected that in the absence of any light, i.e., at zero dose, 50% of random eye shifts will be 'correct' so that the natural response rate is $c = 0.50$. The following data is taken from Barlow and Feigl [1982].

Dose group i	Observations n_i	Intensity (ftl.) dose	$\log_{10}$ x_i	Observations correct r_i^*	% correct p_i^*	Abbott's transforms r_i	p_i
1	40	100	2.000	24	0.6	8	.2
2	40	200	2.301	28	0.7	16	.4
3	40	400	2.602	36	0.9	32	.8

The parameter of interest is the ED'_{75}, the intensity required to produce a 75% correct response ($p^* = 0.75$) which is equivalent to the ED_{50} on the p scale.

Using GLIM3 the best fitting dose-response line by the method of maximum likelihood after two iteration is

$\hat{\eta}$ dose $= -7.086 + 3.025x$, where $\Phi(\eta) = \alpha + \beta x$.

The estimated $ED'_{75} = \text{antilog}_{10} x_0$, where x_0 is the value of x for which $\hat{\eta} = 0$, thus $x_0 = 2.34$, $ED'_{75} = 220$. The 95% confidence intervals for x_0 obtained by Barlow and Feigl [1982] is (2.033, 2.509) and on the original intensity scale it becomes (108, 323). These wide limits can be attributed to the poor design, namely, the poor choice of doses.

Next, towards fitting the logit by the method of least squares we have

$l_i = \ln(p_i/q_i)$, $i = 1, 2, 3 \ldots$, then

$l_i = -1.386$, $l_2 = 0.405$ and $l_3 = -1.386$.

x	n	p	l	$w = np(1-p)$	Wl
2.00	40	0.2	-1.386	6.4	-8.87
2.30	40	0.4	-0.405	9.6	-3.89
2.60	40	0.8	1.386	6.4	8.87

$\Sigma W_i = 22.4$, $\Sigma W_i x_i = 51.54$, $\Sigma W_i x_i^2 = 119.64$,

$\Sigma W_i l_i = 3.89$, $\Sigma W_i l_i x_i = 3.63$.

Hence

$\bar{x} = 51.54/22.4 = 2.30$, $\bar{l} = 3.89/22.4 = -0.17$

$$b = \frac{(22.4)(-3.63) - (-3.89)(51.54)}{(22.4)(119.64) - (51.54)^2} = \frac{119.18}{23.56} = 5.06$$

$a = -0.17 - (5.06)(2.30) = -11.80$

$\log ED'_{75} = -a/b = 2.33$ and $ED'_{75} = 215$.

Also,

$$b^2 SE^2(-a/b) = \frac{1}{22.4} + \left(\frac{0.17}{8.06}\right)^2 [119.64 - (22.4)(2.30)^2]$$

$$= 0.045 + 0.0011 (1.14) = 0.046$$

$SE(-a/b) = 0.042$.

95% confidence limits for $-a/b$ are $2.33 \pm 2(0.042) = (2.245, 2.415)$. Hence, 95% confidence limits for $ED'_{75} = \text{antilog}(2.245, 2.415) = (176, 260)$.

5.3. Estimation of Relative Potency

If the assays of a standard preparation S and a test preparation T are parallel, then the horizontal distance between the lines will be the same and it provides an estimate of the relative potency of T. If the lines are not parallel, then the ratio of the two ED_{50} points on the response curves will provide an estimate of the relative potency. In practice, we shall be comparing only two treatments of similar chemical structure. Hence the lines will be parallel at least in the range of doses of practical interest. Then, if we are fitting a logit model, we will minimize the logit χ^2, given by

$$\sum_{i=1}^{k_1} n_{iT} p_{iT} q_{iT} (l_{i,T} - \hat{l}_{i,T})^2 + \sum_{i=1}^{k_2} n_{iS} p_{iS} q_{iS} (l_{i,T} - \hat{l}_{i,S})^2$$

with respect to the estimates of the three unknown parameters α_T, α_S and β where

$\hat{l}_{i,T} = a_T + bx_i$ and $\hat{l}_{i,S} = a_S + bx_i$,

$l_{i,T} = \alpha_T + \beta x_i$, $l_{i,S} = \alpha_S + \beta x_i$.

The normal equations are

$\Sigma n_{iT} p_{iT} q_{iT} (l_{i,T} - \hat{l}_{i,T}) = 0$

$\Sigma n_{iS} p_{iS} q_{iS} (l_{i,S} - \hat{l}_{i,S}) = 0$ and

$$\sum_{1}^{k_1} n_{iT} p_{iT} q_{iT} x_i (l_{i,T} - \hat{l}_{i,T}) + \sum_{1}^{k_2} n_{iS} p_{iS} q_{iS} x_i (l_{i,S} - \hat{l}_{i,S}) = 0.$$

The horizontal distance between the lines fitted via least squares is

$M = (a_T - a_S)/b$

and the estimate of the relative potency is R, where $\log R = M$. Writing

$\hat{l}_{i,S} = a'_S + (x_i - \bar{x}_S)b$, $a'_S = a_S + b\bar{x}_S$ and

$\hat{l}_{i,T} = a'_T + (x_i - \bar{x}_T)b$, $a'_T = a_T + b\bar{x}_T$

one can show (as in section 4.6) that the estimated variance of M is given by (after noting that $M = \bar{x}_S - \bar{x}_T - (a'_S - a'_T)/b$):

$s_M^2 = [s_{a'_S}^2 + s_{a'_T}^2 + (\bar{x}_S - \bar{x}_T - M)^2 s_b^2]/b^2$.

In carrying out calculations, the appropriate sums of squares and cross-products are pooled in order to obtain b. A test for parallelism can then be made. The sum of squares due to regression can be decomposed into two parts, one part having one degree of freedom for the lines and the other for nonlinearity having $k_1 + k_2 - 4$ degrees of freedom.

If we have two two-point assays, the calculations and fitting of the lines will be much simpler. If the doses are D_1 and $D_2 = rD_1$, the standard and the test preparation are diluted in the same proportion and if the lower and upper levels of each preparation are coded as $x = 0$ and $x = 1$, respectively, then

$\log R = (a_T - a_S)(\log r)/b$.

Example

Decamethonium bromide, known concentration and unknown, is tested by an experiment with 24 animals at each dosage [Berkson, 1953]. It yielded the following data:

Drug	Dose, mg/ml	Designation	x	n	r	p	W	Wl
Tested	D	1	0	24	6	0.250	0.1875	−0.2060
	1.5D	2	1	24	16	0.667	0.2221	0.1543
Standard	0.016	3	0	24	7	0.292	0.2067	−0.1831
	0.024	4	1	24	21	0.875	0.1094	0.2128

Berkson [1953] obtains:

$a_T = -1.3309$

$a_S = -0.6750$

$b = 2.2217$

$M = (a_T - a_S)/b = -0.2952$

$\log R = M \log 1.5 = -0.0520, \quad R = 0.8872$

$s_b^2 = 0.2406, \quad s_b = 0.49$

$\bar{x}_T = 0.5422, \quad \bar{x}_S = 0.3461 \left(\bar{x} = \dfrac{\Sigma W_i x_i}{\Sigma W_i} \right)$

$s_M^2 = 0.04778, \quad s_M = 0.2186$

$s_{\log R} = s_M(\log 1.5) = 0.0385$

$s_R = R \, s_{\log R}/\log e = 0.0786$.

Problem

Dose = number of nematodes/larvae; n = number of test subjects (larvae) per dose; r = number of larvae that died after 5 days; T spec = nematode species; instar = age of the larvae, each instar is proportional to the age of the insect as it appears between molts.
Evaluate the slopes, R and the standard errors.
In this bioassay, 10 insect larvae were exposed to a known number of entomogenous nematodes in a 100 × 15 mm petri dish for 24 h, after which the larvae

were separated and held individually in 1-ounce plastic cups. The mortality was recorded as follows, 5 days after the initial exposure to the nematode:

	1st instar *P. robinie* larvae versus *S. feltial* nematode species					1st instar *P. robinie* larvae versus *S. bibionis* nematode species				
Dose	0.5	1	10	50	200	0.5	1	10	50	200
n	50	50	50	50	30	40	40	40	40	40
r	3	3	25	44	30	1	1	24	36	40

We have the following data (source: Brian Forschler of the Department of Entomology, University of Kentucky):

	3rd instar *P. robinie* larvae versus *S. feltial* nematode species					3rd instar *P. robinie* larvae versus *S. bibionis* nematode species				
Dose	0.5	1	5	10	50	0.5	1	5	10	50
n	30	30	30	30	30	30	30	30	30	30
r	12	19	29	30	30	7	7	22	30	30

	6th instar *P. robinie* larvae versus *S. feltial* nematode species					6th instar *P. robinie* larvae versus *S. bibionis* nematode species					
Dose	0.5	1	5	10	50	0.5	1	5	10	50	200
n	30	130	130	130	130	70	60	70	60	60	25
r	15	53	90	104	120	4	7	18	35	47	24

Note: Problems can be assigned using the above data sets.

5.4. *An Optimal Property of the Logit and Probit Estimates*

Taylor [1953] has shown that a class of estimates which, in particular, includes the logit and probit estimates, is regular and best asymptotically normal (RBAN) in the sense of Neyman [1949]. In the following discussion, we shall specialize all the results from the multinomial response to the Bernoulli response.

There are k sequences of independent trials and each sequence consists of n_i trials. A trial of the i-th sequence is Bernoulli with probabilities P_i and Q_i such that $P_i + Q_i = 1$. Let $N = \sum_{1}^{k} n_i$ and $\lambda_i = n_i/N$. Let θ denote a vector of parameters, $\theta_1, \ldots, \theta_m$ and $P_i = f_i(\theta)$, $Q_i = 1 - f_i$ where f_i are continuous having continuous partial derivatives up to the second order. We assume that $f_i(\theta) > 0$. Further, $\theta_1, \ldots, \theta_m$ are assumed to be functionally independent. That is

$$\begin{vmatrix} f_{j_1,1} & & f_{j_1,m} \\ \vdots & & \\ f_{j_m,1} & & f_{j_m,m} \end{vmatrix} \neq 0 \quad \text{where} \quad f_{i,j} = \frac{\partial f_i}{\partial \theta_j} \quad \text{and} \quad 1 \leq j_1, \ldots, j_m \leq k.$$

Definition. $D(P,P')$ is said to be a distance function of $P = (P_1, \ldots, P_k)$ and $P' = (P'_1, \ldots, P'_k)$ if
(1) $D(P,P) = 0$,
(2) $D(P,P') > 0$ for $P \neq P'$,
(3) $D(P,P')$ is continuous with continuous partial derivatives up to the second order.

Let $p = (p_1, \ldots, p_k)$ denote the observed values of $P = (P_1, \ldots, P_k)$. Let $s_i(p)$ be an estimate of θ_i, $i = 1, \ldots, m$. Let

$\partial D(P,p)/\partial \theta_j = \Psi_j(\theta,p)$.

Lemma (Werner Leimbacher)

Let $D(P,p)$ satisfy:

(4) $\partial^2 D(P,p)/\partial p_i \partial \theta_1|_{p=f} = \partial \Psi_1/\partial p_i|_{p=f} = c\lambda_i f_{i,1}/f_i$

(5) $\partial^2 D(P,p)/\partial \theta_t \partial \theta_r|_{p=f} = \partial \Psi_t/\partial \theta_r|_{p=f} = -c \sum_{i=1}^{k} \lambda_i \frac{f_{it} f_{ir}}{f_i(1 - f_i)}$

where c is a constant. Then the $s_j(p)$ which minimize $D(P,p)$ are RBAN estimates of θ_j ($j = 1, \ldots, m$).

Proof. Verify the sufficient conditions of Neyman [1949, theorem 2, p. 248].

We define a class of distance functions $\mathscr{F}$ as follows.

Definition. $D(P,p)$ is said to belong to $\mathscr{F}$ if it is a distance function and it further satisfies conditions (4) and (5) of the lemma.

Now we are ready to give the main theorem of Taylor.

Taylor's Theorem [1953]

If $h(x)$ is strictly monotone in x for $0 < x < 1$, having continuous derivatives up to the 3rd order, and if the function $g(u,v)$ is positive for $0 < u, v < 1$, has continuous partial derivatives up to the second order, and satisfies the condition

$g(f_i, f_i) = [\partial h/\partial x|_{x=f_i}]^{-2}/f_i$ for all i,

An Optimal Property of the Logit and Probit Estimates

then

$$D_1(P,p) = \sum_{i=1}^{k} n_i\{g(f_i,p_i)[h(p_i) - h(f_i)]^2 + g(1 - f_i, 1 - p_i)[h(1 - p_i) - h(1 - f_i)]^2\}$$

$$= D_{11}(P,p) + D_{12}(P,p)$$

belongs to the class $\mathscr{F}$ (that is, the estimates $s_i(p)$, $i = 1, \ldots, m$ which minimize $D_1(P,p)$ are RBAN estimates).

Proof. $D_1(P,p)$ is a distance function, since $h(x)$ is strictly monotone and continuous and has continuous derivatives. Furthermore, one can easily verify that (since D_{12} is an explicit function of $q_i = 1 - p_i$ and consequently $\partial D_{12}/\partial p_i = 0$, $i = 1, \ldots, k$)

$$\partial^2 D_1/\partial p_i \partial \theta_1 |_{p=f} = -2N\lambda_i \frac{f_{i1}}{f_i}, \quad i = 1, \ldots, k.$$

A similar formula holds for $\partial^2 D_1/\partial p_i \partial \theta_2$ for $i = 1, \ldots, k$. Also we have

$$\partial^2 D_1/\partial \theta_r \partial \theta_t |_{p=f} = 2N \sum_{i=1}^{k} \lambda_i \frac{f_{ir} f_{it}}{f_i(1 - f_i)}.$$

Thus, D_1 belongs to the class $\mathscr{F}$.

Corollary 1

If $h(x) = -h(1 - x)$ for $0 < x < 1$, then $D_1(P,p)$ takes the simple form

$$D_1^*(P,p) = \sum_{i=1}^{k} n_i[g(f_i,p_i) + g(1 - f_i, 1 - p_i)][h(p_i) - h(f_i)]^2.$$

Special Cases

Let $\theta = (\alpha, \beta)$ and assume that

$$\begin{vmatrix} f_{i\alpha} & f_{i\beta} \\ f_{l\alpha} & f_{l\beta} \end{vmatrix} \neq 0 \quad \text{for } 1 \leq i, l \leq k.$$

As a special case we take

$$f_i(\alpha,\beta) = [1 + \exp(-\alpha - \beta x_i)]^{-1} \quad (i = 1, \ldots, k)$$

Consider the logit χ^2

$$\chi_A^2 = \sum_{i=1}^{k} n_i p_i (1 - p_i) \left[\ln \frac{p_i}{1 - p_i} - \ln \frac{f_i}{1 - f_i} \right]^2.$$

Corollary 2

Let $P_i = f_i(\alpha, \beta)$, $i = 1, \ldots, k$ where the f_i are any functions of α and β which are continuous with continuous partial derivatives up to the second order and which satisfy that $0 < P_i < 1$, and there is no functional relationship between α and β. Then χ_A^2 as given above is a member of $\mathscr{F}$.

Proof. Set $h(x) = \ln[x/(1-x)]$ in Corollary 1. Then it satisfies all the conditions and moreover $h(x) = -h(1-x)$. Also

$$\partial h/\partial x|_{x=f_i} = 1/f_i(1-f_i).$$

Obviously

$$p_i(1-p_i) = \frac{p_i^2(1-p_i)^2}{p_i} + \frac{p_i^2(1-p_i)^2}{(1-p_i)}.$$

Also, $g(f_i, p_i) = p_i^2(1-p_i)^2/p_i$ satisfies the regularity conditions imposed by Taylor's theorem. It is positive and has continuous partial derivatives up to the second order and in particular

$$g(f_i, f_i) = \frac{1}{f_i}\left(\frac{\partial h}{\partial x}\bigg|_{x=f_i}\right)^{-2} = f_i^2(1-f_i)^2/f_i.$$

Substituting the expressions for h and g, χ_A^2 takes the form of $D_1^*(P,p)$. Thus, $\hat{\alpha}(p)$ and $\hat{\beta}(p)$ which minimize χ_A^2 will be RBAN estimates of α and β, respectively.

Remark. It is not necessary to write $g(f_i, p_i)$ as a function of two variables when one argument does not appear.

Special Case

Probit estimate

$$P_i = f_i(\mu, \sigma) = \Phi[(x_i - \mu)/\sigma]$$

where Φ denotes the cdf of the standard normal variable. Define

$$\chi_B^2 = \sum_{i=1}^{k} n_i G(f_i, p_i)[\Phi^{-1}(p_i) - \Phi^{-1}(P_i)]^2, \quad \text{where}$$

$$G(f_i, p_i) = [p_i(1-p_i)]^{-1}\left[\frac{\partial \Phi^{-1}(P_i)}{\partial P_i}\bigg|_{P_i=p_i}\right]^{-2}$$

$$= [p_i(1-p_i)]^{-1}\left[\frac{1}{\frac{\partial \Phi(x)}{dx}\bigg|_{x=\Phi^{-1}(p_i)}}\right]^{-2}$$

$$= [p_i(1-p_i)]^{-1}[\phi(\Phi^{-1}(p_i))]^2$$

where $\phi = \dfrac{d}{dx}\Phi(x)$ is the standard normal density. We will find estimates of μ and σ that minimize χ_B^2.

$$\partial \chi_B^2/\partial \mu = \frac{2}{\sigma}\sum_{i=1}^{k} n_i G(f_i, p_i)\left[\Phi^{-1}(p_i) - \frac{x_i - \mu}{\sigma}\right] = 0$$

$$\partial \chi_B^2/\partial \sigma = \frac{2}{\sigma}\sum_{i=1}^{k} n_i G(f_i, p_i)\left[\Phi^{-1}(p_i) - \frac{x_i - \mu}{\sigma}\right]\left(\frac{x_i}{\sigma}\right) = 0.$$

These two equations can be written as

$$\sigma\left(\sum_{i=1}^{k} A_i\right) + \mu\left(\sum_{i=1}^{k} B_i\right) = \sum_{1}^{k} C_i$$

$\sigma(\Sigma A_i x_i) + \mu(\Sigma B_i x_i) = \Sigma C_i x_i$, where

$A_i = \lambda_i G(f_i, p_i) \Phi^{-1}(p_i)$, $B_i = \lambda_i G(f_i, p_i)$

$C_i = \lambda_i G(f_i, p_i) x_i$.

Corollary 3

Let $\chi_B^2 = \Sigma n_i \{[p_i(1-p_i)]^{-1} [\phi(\Phi^{-1}(p_i))]^2\} [\Phi^{-1}(p_i) - \Phi^{-1}(f_i)]^2$

where $f_i(\mu, \sigma) = \Phi[(x_i - \mu)/\sigma]$. Then χ_B^2 is a member of $\mathscr{F}$ and hence it follows that $\hat{\mu}(p)$ and $\hat{\sigma}(p)$, which minimize χ_B^2 are RBAN estimates of μ and σ.

Proof. We apply Corollary 1 and set $h(x) = \Phi^{-1}(x)$ where $\Phi^{-1}(x) = -\Phi^{-1}(1-x)$. Also $G(f_i, p_i)$ can be written as

$p_i^{-1}\{\phi(\Phi^{-1}(p_i))\}^2 + (1-p_i)^{-1}[\phi(\Phi^{-1}(1-p_i))]^2$

$= g(f_i, p_i) + g(1-f_i, 1-p_i)$

where $g(u, v)$ satisfies the conditions of Taylor's theorem. Hence, χ_B^2 can be put in the form of $D_i^*(P, p)$ and thus χ_B^2 belongs to $\mathscr{F}$.

Concluding Remarks. If the n_i are sufficiently large such that λ_i are held constant, then the above method of estimating the parameters (α, β) or (μ, σ) is better than the lengthy iterative process of maximum likelihood estimation.

5.5. *Confidence Bands for the Logistic Response Curve*

One of the problems of interest in logit analysis is the interval estimation of the logistic response curve. Let Y denote the response variable and let $X_1, \ldots, X_k$ denote explanatory variables. We assume that Y is a dichotomous variable taking the values 0 or 1. Let the expectation of Y, namely, P(X), be related to the set of explanatory variables, $X_1, \ldots, X_k$ by

$P(X) = \exp[\lambda(X)]/\{1 + \exp[\lambda(X)]\}$ where (1)

$X = (1, X_1, \ldots, X_k)'$

$\lambda(X) = \beta'X$, $\beta = (\beta_0, \ldots, \beta_k)'$.

Alternatively, one can rewrite the model (1) in terms of P(X) as

$\lambda(X) = \ln\{P(X)/[1 - P(X)]\}$.

The unknown parameters β_i are estimated from the random sample of size N

$(X_1, Y_1) \ldots (X_N, Y_N)$.

When the number of explanatory variables is one (that is k = 1), Brand et al. [1973] give a method for obtaining a confidence band for the logistic response function based on the large sample distribution of the maximum likelihood estimators of the logit model parameters. Hauck [1983] gives an alternative method of obtaining the logistic confidence band which follows closely the corresponding method for regression models. The latter method is computationally easier than the method of Brand et al. [1973]. In the following we present the method of Hauck [1983].

We assume that β is estimated using the method of maximum likelihood assuming that the X values are fixed and that the assumptions on the sequence $X_1, \ldots, X_N$, necessary for asymptotic normality of the maximum likelihood estimators are satisfied (see, for instance, Bradley and Gart [1962]). Then, for sufficiently large N, $N^{1/2}(\hat{\beta} - \beta)$ is (k + 1)-variate normal with mean 0 and variance-covariance matrix Σ. Using this result, it follows that $N(\hat{\beta} - \beta)'\Sigma^{-1}(\hat{\beta} - \beta)$ is asymptotically χ^2 with (k + 1) degrees of freedom. One can estimate the variance-covariance matrix of $\hat{\beta}$, namely, Σ/N by J^{-1} where the (i,j) element of J is given by

$$J_{ij} = -\frac{\partial^2}{\partial \beta_i \partial \beta_j} l(\beta|X_1,\ldots,X_N)|_{\beta=\hat{\beta}}, \quad i, j = 0, \ldots, k. \tag{2}$$

and $l(\beta|X_1,\ldots,X_N)$ is the natural logarithm of the likelihood of β. Thus, using (2) we have for large N

$$P[(\hat{\beta} - \beta)'J(\hat{\beta} - \beta) \leq \chi^2_{k+1, 1-\alpha}] = 1 - \alpha \tag{3}$$

which defines the $1 - \alpha$ confidence ellipsoid for β. However, from Rao [1965, p. 43] it follows that for any fixed vector U and positive definite matrix A:

$$U'A^{-1}U = \sup_X \frac{(U'X)^2}{X'AX} \geq \frac{(U'X)^2}{X'AX}. \tag{4}$$

Now, letting $U = (\hat{\beta} - \beta)$ and $A^{-1} = J$, we obtain

$$(\hat{\beta} - \beta)'J(\hat{\beta} - \beta) \geq [(\hat{\beta} - \beta)'X]^2/[X'JX], \quad \text{for any X.} \tag{5}$$

Using (5) in (3) we have

$$1 - \alpha \leq P\{[(\hat{\beta} - \beta)'X]^2/[X'J^{-1}X] \leq \chi^2_{k+1, 1-\alpha}, \quad \text{for all X}\}$$
$$= P[|(\hat{\beta} - \beta)'X| \leq (\chi^2_{k+1, 1-\alpha}X'J^{-1}X)^{1/2} \quad \text{for all X}].$$

Thus, a conservative $1 - \alpha$ confidence band for $\lambda(X) = \beta'X$ is

$$[\hat{\lambda}_L(x), \hat{\lambda}_U(x)] = [\hat{\beta}'X - (\chi^2_{k+1, 1-\alpha}X'J^{-1}X)^{1/2}, \hat{\beta}'X + (\chi^2_{k+1, 1-\alpha}X'J^{-1}X)^{1/2}]. \tag{6}$$

The corresponding confidence band for P(X) is given by taking the inverse logit transform of (6):

$$[\hat{P}_L(X), \hat{P}_U(X)] = \{e^{\hat{\lambda}_L(X)}/[1 + e^{\hat{\lambda}_L(X)}], e^{\hat{\lambda}_U(X)}/[1 + e^{\hat{\lambda}_U(X)}]\}. \tag{7}$$

Example

Hauck [1983] applies his formula to the data from the Ontario Exercise Collaborative Study [Rechnitzer et al., 1975, 1982]. Here the explanatory variables are

X_1 = smoking status taking values 0 or 1

X_2 = serum triglyceride level $- 100.00$

$N = 341$.

The logit analysis results are

$\hat{\beta}' = (-2.2791, 0.7682, 0.001952)$ and

$$J^{-1} = \begin{vmatrix} -0.06511 & & \\ -0.04828 & 0.09839 & \\ -0.0001915 & 0.00003572 & 0.000002586 \end{vmatrix}.$$

The authors plots the two sets of confidence bands 50 and 95% for smokers ($X_1 = 1$) and for nonsmokers ($X_1 = 0$). The effect of smoking status is clearly shown as is the large uncertainty in the response curves at the extremes of the observed triglycride values.

5.6. Weighted Least Squares Approach

Here, we provide a noniterative procedure for estimating the unknown parameters α and β.

Let $P_i = F(\alpha + \beta x_i)$, $i = 1, 2, \ldots, k$. Let n_i subjects be administered the dose x_i, and r_i denote the number responding at x_i. Also, let

$p_i = r_i/n_i$ and $l_i = G(p_i)$, $G(p) = F^{-1}(p)$ since

$dl = dp/g(p)$, where $g(y) = G'(p)$

$\text{var } l \doteq \text{var } p/(g^2(p))$. Hence

$s_i^2 \doteq p(1-p)/ng^2(p) = 1/W(p)$ (say).

For instance, in the logistic case

$G(y) = \ln[y/(1-y)]$ and $g(y) = [y(1-y)]^{-1}$. Hence

$W(p) = np(1-p)$.

In the probit case $G(y) = \Phi^{-1}(y)$ and $g(y) = 1/\varphi[\Phi^{-1}(y)]$, and consequently $W(p) = n\varphi^2[\Phi^{-1}(p)]/p(1-p) = n\varphi^2(l)/p(1-p)$. Now, by the method of weighted least squares, we wish to find a and b for which

$$\sum_{i=1}^{k} W_i(l_i - a - bx_i)^2, \quad W_i = W(p_i)$$

is minimized.

Let $\bar{x} = \Sigma W_i x_i / \Sigma W_i$, $\bar{l} = \Sigma l_i W_i / \Sigma W_i$. Then

$$b = \Sigma W_i(x_i - \bar{x})(l_i - \bar{l}) / \Sigma W_i(x_i - \bar{x})^2$$

$$= \frac{(\Sigma W_i)(\Sigma W_i x_i l_i) - (\Sigma W_i l_i)(\Sigma W_i x_i)}{(\Sigma W_i)(\Sigma W_i x_i^2) - (\Sigma W_i x_i)^2} \quad \text{and}$$

$a = \bar{l} - b\bar{x}$.

The estimate of $\log LD_{50}$ is $-a/b$ and

$$b^2 SE^2(-a/b) = (\Sigma W_i)^{-1} + (\bar{l}/b)^2 [\Sigma W_i(x_i - \bar{x})^2]^{-1}.$$

Using the above expression, one can set up confidence intervals for $\log LD_{50}$.

Remark. Brown [1970] points out that in the logit case, the least squares estimator of $\log LD_{50}$ will be fully efficient for large samples and will be reasonable for small samples; Monte Carlo studies, carried out with squared error as the criterion, indicate that a and b are as good as the corresponding mles. However, $-a/b$ seem to be slightly interior to its mle.

Let us apply the method of least squares to the assay of estrone by Biggers [1950] in 50% aqueous glycerol by the intravaginal route. The response is the appearance of cornified epithelial cells in vaginal smears taken at specified time after estrogen administration.

Dose 10^{-3} mg	$\log_{10}$ dose, x	Number of mice, n	Proportion of mice responding, p	$l = \ln[p(1-p)]$	$W = np(1-p)$
0.2	−0.7	27	0.15	−1.73	3.44
0.4	−0.4	30	0.43	−0.28	7.35
0.8	−0.1	30	0.60	0.41	7.20
1.6	0.2	30	0.73	0.99	5.91

Here $\Sigma W_i = 23.90$, $\Sigma W_i x_i = -4.89$, $\Sigma W_i x_i^2 = 3.17$

$\Sigma W_i l_i = 0.79$, $\Sigma W_i x_i l_i = 5.87$

$$\bar{x} = -0.20, \quad \bar{l} = 0.03, \quad b = \frac{(23.9)(5.87) - (-4.89)(0.79)}{(23.9)(3.17) - (-4.89)^2} = \frac{144.156}{51.85} = 2.78$$

$a = 0.03 - (2.78)(-0.20) = 0.59$

$\log LD_{50} = -0.21$, $LD_{50} = 0.62$

$b^2 SE^2(-a/b) = 0.04 + 0.00005 = 0.04$. Hence, $SE(-a/b) = 0.07$.

95% limits are antilog $[\log LD_{50} \pm 2 SE(-a/b)]$

$$= \text{antilog}(-0.35, -0.07) = (0.45, 0.85).$$

6 The Angular Response Curve

Wilson and Worcester [1943d] have considered the bioassay on a general curve and in particular, the angular response curve.

6.1. *Estimation by the Method of Maximum Likelihood*

Let the response to dose level x be given by the general function, namely

$$P = [1 + F(\alpha + \beta x)]/2 = [1 + F(y)]/2 \qquad (6.1)$$

where F is some given function and $y = \alpha + \beta x$. For the logistic (or growth curve) $F(y) = \tanh(y/2)$ and for the angular transformation $F(y) = \sin y$, $-\pi/2 \leqslant y \leqslant \pi/2$. Assuming that at dose level x_i, r_i experimental units out of n_i respond to the dose $(i = 1, \ldots, k)$ where r_i has a binomial distribution with parameters n_i and P_i, the log likelihood of α and β is

$$L = \Sigma \ln C_i + \Sigma r_i \ln(1 + F_i) + \Sigma(n_i - r_i)\ln(1 - F_i) - N\ln 2 \qquad (6.2)$$

where $C_i = \binom{n_i}{r_i}$, $F_i(x) = F(\alpha + \beta x_i)$ and $N = n_1 + \cdots + n_k$. Letting $f = F' = \partial F/\partial y$ and $f' = \partial f/\partial y$ and noting that $\partial y/\partial \alpha = 1$ and $\partial y/\partial \beta = x$ we have

$$\begin{aligned}\partial L/\partial \alpha &= \Sigma 2 r_i f_i (1 - F_i^2)^{-1} - \Sigma n_i f_i (1 - F_i)^{-1} = 0 \\ \partial L/\partial \beta &= \Sigma 2 r_i x_i f_i (1 - F_i^2)^{-1} - \Sigma n_i x_i f_i (1 - F_i)^{-1} = 0.\end{aligned} \qquad (6.3)$$

The above equations can be explicitly solved only when the number of dose levels is two, in which case we have, from the definition of F,

$$F^{-1}(2P_i - 1) = \alpha + \beta x_i \quad (i = 1,2).$$

Consequently,

$$\beta = [F^{-1}(2P_2 - 1) - F^{-1}(2P_1 - 1)]/(x_2 - x_1) \quad \text{and}$$

$$\begin{aligned}\alpha &= F^{-1}(2P_1 - 1) - x_1(x_2 - x_1)^{-1}[F^{-1}(2P_2 - 1) - F^{-1}(2P_1 - 1)] \\ &= [x_2 F^{-1}(2P_1 - 1) - x_1 F^{-1}(2P_2 - 1)]/(x_2 - x_1).\end{aligned}$$

Now, estimates a and b for α and β, respectively can be obtained by replacing P_i by their estimates $p_i = r_i/n_i$ $(i = 1,2)$ in the above equations. Special cases can be obtained by setting $F^{-1}(u) = 2\tanh^{-1}u$ and $F^{-1}(u) = \sin^{-1}u$ for the logistic and angular cases, respectively.

The mle values of α and β can be obtained from (6.3) by successive iterations. We need the matrix of second derivatives

$$\partial^2 L/\partial\alpha^2 = \Sigma 2r_i f_i'(1-F_i^2)^{-1} + \Sigma 4r_i f_i^2 F_i (1-F_i^2)^{-2}$$
$$- \Sigma n_i f_i'(1-F_i)^{-1} - \Sigma n_i f_i^2 (1-F_i)^{-2}.$$

After noting that $Er_i = (1+F_i)n_i/2$, we have

$$E(\partial^2 L/\partial\alpha^2) = -\Sigma n_i f_i^2 (1-F_i^2)^{-1}.$$

Similarly, one can obtain

$$E(\partial^2 L/\partial\alpha\partial\beta) = -\Sigma n_i x_i f_i^2 (1-F_i^2)^{-1} \quad \text{and}$$
$$E(\partial^2 L/\partial\beta^2) = -\Sigma n_i x_i^2 f_i^2 (1-F_i^2)^{-1}.$$

Let $D =$ det. of the matrix of the expectation of second derivatives. Let

$$s_i = F^{-1}(2p_i - 1), \quad p_i = r_i/n_i \quad (i=1,\ldots).$$

One can fit a provisional line for the set of points (x_i, s_i) $i=1,\ldots,k$ and let a_0, b_0 be the initial values of the intercept and slope. Let $s_{i0} = F^{-1}(2\hat{P}_{i,0} - 1)$ where $\hat{P}_{i,0}$ corresponds to the ordinate at x_i on the provisional line. $\partial L/\partial\alpha = 0$ implies that

$$\Sigma n_i(p_i - \hat{P}_i)f_i(1-F_i^2)^{-1} = 0 \quad [F_i = F(\hat{s}_i) = 2\hat{P}_i - 1, \; \hat{s}_i = a + bx_i]$$

and $\partial L/\partial\beta = 0$ implies that

$$\Sigma n_i(p_i - \hat{P}_i)x_i f_i(1-F_i^2)^{-1} = 0.$$

For the angular transformation $f(s)/[1-F^2(s)] = \sec.s$. Also write

$$p_i - \hat{P}_i = p_i - \hat{P}_{i0} + \hat{P}_{i0} - \hat{P}_i \quad \text{and}$$
$$\hat{P}_{i0} - \hat{P}_i = \tfrac{1}{2}[F(s_{i0}) - F(\hat{s}_i)] \doteq \tfrac{1}{2}(s_{i0} - \hat{s}_i)f(s_{i0}) = -\tfrac{1}{2}(\delta a + x_i \delta b)f(s_{i0});$$

substituting these relations in the likelihood equations, we obtain, since $F(s) = \sin s$,

$$\Sigma n_i(p_i - \hat{P}_{i0})\sec s_{i0} = \tfrac{1}{2}\Sigma n_i(\delta a + x_i\delta b),$$
$$\Sigma n_i(p_i - \hat{P}_{i0})x_i \sec s_{i0} = \tfrac{1}{2}\Sigma n_i(\delta a + x_i\delta b)x_i$$

where $s_{i0} = a_0 + b_0 x_i$. Thus,

$$\begin{vmatrix} \Sigma n_i & \Sigma n_i x_i \\ \Sigma n_i x_i & \Sigma n_i x_i^2 \end{vmatrix} \begin{vmatrix} \delta a \\ \delta b \end{vmatrix} = \begin{vmatrix} 2\Sigma n_i(p_i - \hat{P}_{i0})\sec s_{i0} \\ 2\Sigma n_i(p_i - \hat{P}_{i0})x_i \sec s_{i0} \end{vmatrix} \quad \text{or}$$

$$\begin{vmatrix} \delta a \\ \delta b \end{vmatrix} = \frac{1}{D}\begin{vmatrix} \Sigma n_i x_i^2 & -\Sigma n_i x_i \\ & \Sigma n_i \end{vmatrix}\begin{vmatrix} 2\Sigma n_i(p_i - \hat{P}_{i0})\sec s_{i0} \\ 2\Sigma n_i(p_i - \hat{P}_{i0})x_i \sec s_{i0} \end{vmatrix}.$$

where $D = (\Sigma n_i)[\Sigma n_i(x_i - \bar{x})^2]$, $\bar{x} = \Sigma n_i x_i/\Sigma n_i$.

Notice that iteration here should be easy since the inverse of the information matrix can be evaluated once and for all, and only the right side vector needs to be evaluated at each iteration. Since $\hat{P} = \frac{1}{2}(1 + \sin s)$, $1 - \hat{P} = \hat{Q} = \frac{1}{2}(1 - \sin s)$, $\sec s = \frac{1}{2}(\hat{P}\hat{Q})^{-1/2}$. Thus $\sec s_{i0} = \frac{1}{2}(\hat{P}_{i0}\hat{Q}_{i0})^{-1/2}$.

6.2. Alternative Method of Estimation

First-order Taylor's expansion gives

$$p_i - \hat{P}_i = (s_i - \hat{s}_i)(\hat{P}_i\hat{Q}_i)^{1/2}$$

Substituting this in the likelihood equations we readily obtain

$$\Sigma n_i(s_i - \hat{s}_i) = 0, \quad \text{and}$$

$$\Sigma n_i x_i(s_i - \hat{s}_i) = 0, \quad \text{where } \hat{s}_i = a + bx_i$$

Hence

$$b = \Sigma n_i(x_i - \bar{x})(s_i - \bar{s}) / \Sigma n_i(x_i - \bar{x})^2$$

$$a = \bar{s} - b\bar{x}, \quad \text{where}$$

$$\bar{x} = \Sigma n_i x_i / N, \quad \bar{s} = \Sigma n_i s_i / N, \quad N = \Sigma n_i.$$

Notice that the latter method does not require iteration. In either case

$$\sigma_b^2 = [\Sigma n_i(x_i - \bar{x})^2]^{-1}, \quad \text{since } \text{var } s_i \doteq n_i^{-1} \quad (i = 1, \ldots, k)$$

(which follows from the method of differentials: $\cos s_i\, ds_i = 2\, dp_i$).

$$\sigma_a^2 = \frac{1}{N} + \frac{\bar{x}^2}{\Sigma n_i(x_i - \bar{x})^2} \quad \text{and}$$

$$\text{cov}(a, b) = -\Sigma n_i x_i / \{N \Sigma n_i(x_i - \bar{x})^2\}.$$

Hence, an estimate of the variance of $c = -a/b$ is

$$s_c^2 = b^{-2}[\text{var } b + c^2 \text{var } a - 2c\, \text{cov}(a, b)]$$

$$= b^{-2} s_{a'}^2 + (c - \bar{x})^2 s_b^2, \quad a' = a + b\bar{x}$$

$$= \left[\frac{1}{N} + \frac{(c - \bar{x})^2}{\Sigma n_i(x_i - \bar{x})^2}\right] b^{-2}.$$

6.3. Comparison of Various Methods

There are three methods that are currently used in practice: (1) the probit method, (2) the logit method and (3) the angular transformation method. In the range of interest for the dosage, they all look alike and it does not matter which method you use. It seems to be a matter of taste. The logit method seems to be natural and easy to use. The minimum logit method is elegant. It is possible that one could vary the linear effect function from $\alpha + \beta x$ to $\alpha + \beta x + \gamma x^2$ or to a polynomial function.

6.4. Other Models

Other functions considered in the literature are the rectangular and the one suggested by Wilson and Worcester [1943b]

$dP = \beta f(\alpha + \beta x) dx$, where

$f(y) = \frac{1}{2}$ ($-1 < y < 1$, and zero elsewhere)

or $f(y) = \frac{1}{2}(1 + y^2)^{-3/2}$ ($-\infty < y < \infty$).

The latter implies that

$$P(y) = \left(\frac{1}{2}\right)\left[1 + \frac{y}{(1+y^2)^{1/2}}\right].$$

Some investigators, for instance, Bartlett [1936] and Bliss [1937] suggested the transformation $P = \sin^2\theta$ ($0 \leq \theta \leq \pi/2$).

Since $dp = 2\sin\theta\cos\theta\, d\theta$ where p denotes the sample proportion responding to the treatment,

$$\operatorname{var}\hat{\theta} \approx \frac{\operatorname{var} p}{4\sin^2\theta\cos^2\theta} = \frac{P(1-P)}{4nP(1-P)} = \frac{1}{4n}, \quad \text{if } \theta \text{ is expressed in radians,}$$

$$= \left(\frac{180}{\pi}\right)^2 \frac{1}{4n} = \frac{821}{n}, \quad \text{if } \theta \text{ is expressed in degrees.}$$

When the above transformation is used in the bioassay problems, $\hat{\theta}$ as an estimate of $\alpha + \beta x$ is biased. Anscombe [1956] has suggested the following modification in order to overcome this bias. He suggests to get

$$\hat{\theta} = \sin^{-1}\left\{\left(\frac{r + 1/4}{n + 1/2}\right)^{1/2}\right\}, \quad \text{instead of } \hat{\theta} = \sin^{-1}\{(r/n)^{1/2}\}.$$

Then, one can show (after expanding the right side in powers of nP)

$E\hat{\theta} = \alpha + \beta x + 0(n^{-2})$, $\operatorname{var}\hat{\theta} \cong (4n)^{-1}$

$\beta_1^{1/2}(\hat{\theta}) \simeq (P - Q)/2(nPQ)^{1/2}$

where $\beta_1(\theta)$ denotes the skewness.

6.5. *Comparison of Maximum Likelihood (ML) and Minimum χ^2 Estimates (MCS)*

For some simple logit [Berkson, 1955] and probit [Berkson, 1957a] models, the MCS estimates had smaller mean square errors (MSE) than the ML estimates.

Approximations to the MSEs of the two estimates to the order n^{-2} (where n denotes the average number of observations per cell) were derived by Amemiya [1980] for the logit model and found that the second-order approximation to the MSE of the MCS estimates was smaller than the corresponding ML approximation in every example presented. Although the MCS estimates seem to be superior to the ML estimates with respect to the quadratic loss function, there is no theoretical justification. However, it should be noted that asymptotically MCS and ML estimates are equivalent.

Smith et al. [1984] have carried out Monte Carlo studies in order to study the small-sample behavior of maximum likelihood and minimum χ^2 estimates in a simple dichotomous logit regression model. Statistics designed to test regression coefficients converge slowly to normality for designs in which doses are placed asymmetrically about the ED_{50}. Designs with doses symmetric about the ED_{50} can be used with confidence at moderate sample sizes. There is some evidence that the ML is preferable to MCS when statistical inferences are to be made with the logit model.

6.6. More on Probit Analysis

Fitting a probit regression line by eye to the results of special species of nematodes.

Table 6.1. Data of 6th instar *P. robiniae* larvae versus *S. bibionis* nematode species (source: Brian Forschler of the Department of Entomology, University of Kentucky)

Dose	0	0.5	1	5	10	50	200
log dose		−0.30	0	0.70	1	1.70	2.30
n	70	70	60	70	60	60	25
r (died)	0	4	7	18	35	47	24
p (% killed)	0	5.7	11.7	25.7	58.3	78.3	96.0
Empirical probits		−1.58	−1.18	−0.75	0.21	0.78	1.75
Expected probits		−1.48	−1.14	−0.34	0.00	0.80	1.48

The empirical probits are obtained as

$\Phi^{-1}(p_i) = (x_i - \mu)/\sigma, \quad x_i = \log \text{dose}_i \quad (i = 1,\ldots,6).$

The first column suggests that the natural mortality of the *P. robiniae* larvae is zero. Hence there is no need to make any adjustment for natural mortality. When the empirical probits are plotted against x_i, they lie nearly on a straight line and such a line can be drawn by eye. From this line probits corresponding to different values of x are read and are given in the last row of table 6.1. They are converted back to percentages by the standard normal table. From figure 1 we see that a probit value of 0.0 corresponds to a dose level m = 1.0 which is the estimate of log LD_{50}. Thus, LD_{50} is estimated as 10.0. In order to estimate the slope, we note that a change of 0.7 in x will result in an increase of 0.8 in the probit. Thus, the estimated regression coefficient of probit on log dose, or the rate of increase of probit per unit increase in x is

b = 0.8/0.7 = 1.14.

Also

b = 1/s

where s = $\hat{\sigma}$ is the estimate of σ = 1/1.14 = 0.875.
Thus, the relation between probit and log dose is

y = 1.14(x − 1.0).

Now we compute the expected probits using the above linear relation and test the goodness of the model.

x	y	P = Φ^{-1}(y)	n	r	nP	r − nP	(r − nP)^2/nP(1 − P)
−0.30	−1.482	0.069	70	4	4.83	−0.83	0.15
0	−1.140	0.127	60	7	7.62	−0.62	0.06
0.70	−0.342	0.367	70	18	25.69	−7.69	3.64
1.0	0.00	0.500	60	35	30.00	5.00	1.67
1.70	0.798	0.787	60	47	42.22	4.78	2.54
2.30	1.482	0.931	25	24	23.28	0.72	0.32

$$\chi_4^2 = \sum_{i=1}^{6} (r_i - nP_i)^2/nP_i(1 - P_i) = 8.38$$

P-value = $P(\chi_4^2 > 8.38) \doteq 0.08$.

Hence the probit regression line is adequate. Notice that the number of degrees of freedom is 4 since we have estimated two parameters, namely μ and σ. Recall that for the probit

$$P = [\sigma(2\pi)^{1/2}]^{-1} \int_{-\infty}^{x} \exp[-(t - \mu)^2/2\sigma^2]\,dt.$$

If

$$y = \alpha + \beta x = \beta(x + \alpha/\beta),$$

then

$\mu = -\alpha/\beta$ and $\sigma = 1/\beta$.

Hence,

$\partial P(y)/\partial y = \phi(y)$, $\partial P/\partial \alpha = \phi(y)$ and $\partial P/\partial \beta = x\phi(y)$.

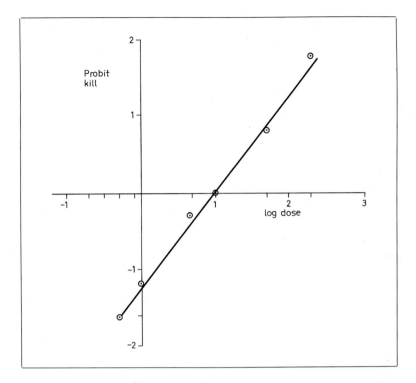

Fig. 1

Example 1

Data taken from Rao [1954].
Artificial data on two response measurements (y_1, y_2):

Standard preparation, IU						Test preparation, mg					
1.25		2.50		5.00		0.125		0.250		0.500	
y_1	y_2	y_1	y_2	y_1	y_2	y_1	y_2	y_1	y_2	y_1	y_2
38	51	53	49	85	47	28	53	48	48	60	43
39	55	102	53	144	51	65	53	47	51	130	50
48	46	81	46	54	39	35	52	54	48	83	48
62	51	75	51	85	41	36	54	74	50	60	51
187	203	311	199	368	178	164	212	223	197	333	192

$\Sigma y_{1s} = 866$, $\Sigma y_{2s} = 580$ $\Sigma y_{1t} = 720$, $\Sigma y_{2t} = 601$

(1) Test for parallelism.
(2) Test for linearity of regression.
(3) Test for significance of the common regression coefficient.

(4) (a) Test whether the extra variable y_2 gives any further information for the estimation of relative potency.
(b) Test whether the estimates of the relative potency from different individual response measurements are the same.
(c) Obtain confidence bounds for the relative potency.
Between sum of squares for

$$y_1 = \frac{(187)^2}{4} + \frac{(311)^2}{4} + \cdots + \frac{(333)^2}{4} - \frac{(1586)^2}{24};$$

Model:
Standard $y = \beta'x + \varepsilon$, $\beta' = (\beta_1, \beta_2)$
Test $y = \beta'x + \varepsilon$.

for

$$y_2 = \frac{1}{4}[(203)^2 + (199)^2 + \cdots + (192)^2] - \frac{(1181)^2}{24}.$$

The doses can be coded such that they are $-1, 0, 1$ of the standard and test preparations. For further details see Rao [1954, pp. 208–220].

6.7. *Linear Logistic Model in 2 × 2 Contingency Tables*[1]

Let

$$P_i = [1 + \exp(-x_i'\beta)]^{-1}, \quad \text{and}$$

$$\lambda_i = \ln\left(\frac{P_i}{1-P_i}\right),$$

where $x_i = (x_{i1}, \ldots, x_{is})$ and $\beta = (\beta_1, \ldots, \beta_s)$, $s \geq 2$.

Consider a 2×2 contingency table in which there are two different values for the probability of success, namely P_1 and P_2 corresponding to the two groups of observations. The associated linear logistic model is

$$\lambda_1 = \alpha \quad \lambda_2 = \alpha + \Delta$$

where Δ denotes the difference between the two groups in a logistic scale

$$\Delta = \lambda_1 - \lambda_2 = \log\left[\frac{P_2(1-P_1)}{P_1(1-P_2)}\right].$$

$\exp(\Delta)$ is the ratio of the odds of success versus failure in the two groups. Among all criteria for studying the difference between groups, the measure Δ seems to be valid under a broad range of configuration of the parameters P_1 and P_2 than $\delta = P_2 - P_1$. This is clear because with a fixed Δ, α can vary arbitrarily so that we have a class of situations in which the overall proportion of success is arbitrary; however, a specified difference δ is consistent with only a limited range of individual values for P_1 and P_2. It should be noted that Berkson [1958], Sheps [1958, 1981] and Feinstein [1973] are critical of taking the ratio of the

---

[1] Cox [1970] served as a source for parts of the material in sections 6.7 and 6.8.

rates as a measure of association. They point out that the level of the rates is lost. For further details see, for instance, Fleiss [1981, chapter 6].

Let us consider the following case. For each member of a population, there are two binary variables U and W. Thus, there are four types of individuals corresponding to $U = W = 0$, $U = 1$, $W = 0$; $W = 1$, $U = 0$; $U = W = 1$, the corresponding probabilities being $\pi_{00}, \pi_{10}, \pi_{01}, \pi_{11}$, respectively. For instance,

$$U = \begin{cases} 1 & \text{if he is a nonsmoker} \\ 0 & \text{if he is a smoker} \end{cases} \quad W = \begin{cases} 1 & \text{if he has cancer} \\ 0 & \text{if he has no cancer.} \end{cases}$$

Suppose that we are interested in comparing the probabilities that $W = 1$ for two groups of individuals for which $U = 0$ and $U = 1$. There are at least three sampling procedures by which data can be generated in order to study this problem.

(1) First a random sample may be obtained from the entire population, which enables us to estimate any function of the π_{ij} where π_{ij} is the probability of the (i,j)-th cell. In particular

$P(W = 1 | U = 0) = \pi_{01}/(\pi_{00} + \pi_{01})$ and

$P(W = 1 | U = 1) = \pi_{11}/(\pi_{10} + \pi_{11})$

and the difference of the logistic transforms of these is

$$\log\left(\frac{\pi_{11}}{\pi_{01}}\right) - \log\left(\frac{\pi_{10}}{\pi_{00}}\right) = \log\left(\frac{\pi_{11}\pi_{00}}{\pi_{10}\pi_{01}}\right) = \log\left(\frac{P(W|U)}{P(W|\bar{U})}\right) - \log\left(\frac{P(\bar{W}|U)}{P(\bar{W}|\bar{U})}\right)$$

if $U = 0 [W = 0]$ is denoted by $\bar{U}[\bar{W}]$.

(2) We can draw two random samples, one from each of the subpopulations for which $U = 0$ and $U = 1$, respectively, the sample sizes being fixed and having no relation to the marginal probabilities $\pi_{00} + \pi_{01}$ and $\pi_{10} + \pi_{11}$. Suppose we take equal sample sizes and the values of W obtained in a prospective study (in which one of the populations is defined by the presence and the second by the absence of a suspected previous factor). Individual π_{ij} values cannot be estimated and we can estimate only functions of the conditional probabilities of W given $U = 0$ and $U = 1$: In particular, the logistic difference given above.

(3) Thirdly, we can draw random samples from subpopulations for which $W = 0$ and $W = 1$. Thus, we might take equal subsample sizes, constituting a retrospective study (in which one of the two populations is defined by the presence and the second by the absence of the outcome under study). Here we find the value of U for each patient who has cancer and who does not have cancer. In this study we can estimate functions of

$\pi_{10}/(\pi_{00} + \pi_{10})$ and $\pi_{11}/(\pi_{01} + \pi_{11})$.

Now the logistic difference of these is the same as obtained in scheme (1). Thus, the logistic difference is the same for both the prospective and retrospective studies. Instead, if we wish to estimate the difference

$$P(W = 1|U = 0) - P(W = 1|U = 1) = \pi_{01}(\pi_{00} + \pi_{01})^{-1} - \pi_{11}(\pi_{10} + \pi_{11})^{-1},$$

this can be done from scheme (1) and not from scheme (3).

Let R_j respond to dose level j among n_j subjects where the probability of responding is P_j. Then recall that we noted from Anscombe [1956] that

$$Z_j = \log\left[\left(R_j + \frac{1}{2}\right)\bigg/\left(n_j - R_j + \frac{1}{2}\right)\right]$$

is nearly unbiased for

$\lambda_j = \ln[P_j/(1 - P_j)]$ and

$V_j = (n_j + 1)(n_j + 2)/n_j(R_j + 1)(n_j - R_j + 1)$

is also nearly unbiased for the variance of Z_j [Gart and Zweifel, 1967].

In a 2×2 contingency table, $Z_2 - Z_1$ is an estimate of the logistic difference Δ and has a standard error of approximately $(V_1 + V_2)^{1/2}$. Hence, using the approximate normality of the distribution of $Z_2 - Z_1$, one can test H_0 ($\Delta = 0$) and set up confidence intervals for Δ. Since the logistic model is saturated (i.e. there are as many independent logistic parameters as there are independent binomial probabilities) the unweighted combination $Z_2 - Z_1$ is the unique estimate of Δ from the least-squares analysis.

Example 2

Data from Sokal and Rohlf [1981].

A sample of 111 mice was divided into two groups: 59 that received a standard dose of pathogenic bacteria followed by an antiserum and a control group of 54 that received the bacteria but no antiserum. After the incubation period and the time for the disease to run its due course had elapsed, it was found that 38 mice were dead and 78 survived the disease. Among the mice that died, 13 had received bacteria and antiserum while 25 had received bacteria only. It is of interest to find out whether the antiserum had in any way protected the mice. The data is displayed in the following 2×2 table:

	Bacteria and antiserum	Bacteria only	Total
Dead	13	25	38
Alive	44	29	73
Total	57	54	111

Proportion of those rats that received bacteria and antiserum = 0.514; proportion of those rats that received bacteria only = 0.486.

$Z_1 = \ln(44/13) = 1.219$, $V_1 = 0.095$

$Z_2 = \ln(29/25) = 0.148$, $V_2 = 0.073$

$Z_1 - Z_2 = 1.071$ which has mean Δ and standard deviation, $(0.095 + 0.073)^{1/2} = 0.41$.

We reject H_0 ($\Delta = 0$) against H_1 ($\Delta \neq 0$) at any significance level > 0.009. The 95% confidence interval for Δ is (0.267, 1.875).

6.8. Comparison of Several 2 × 2 Contingency Models

If we wish to combine the information from several 2×2 tables by a weighted mean, then Cox [1970, pp. 78–79] suggests that we use the weighted transforms given by

$$Z_j^{(W)} = \log\left(\frac{R_j - \tfrac{1}{2}}{n_j - R_j - \tfrac{1}{2}}\right) \quad \text{and} \quad V_j^{(W)} = (n_j - 1)/R_j(n_j - R_j).$$

For the j-th 2×2 configuration, we consider

$$\tilde{\Delta}_j^{(W)} = \log\left(\frac{R_{j2} - \tfrac{1}{2}}{n_{2j} - R_{j2} - \tfrac{1}{2}}\right) - \log\left(\frac{R_{j1} - \tfrac{1}{2}}{n_{j1} - R_{j1} - \tfrac{1}{2}}\right)$$

with an associated variance

$$V_{\Delta_j}^{(W)} = \frac{n_{j2} - 1}{R_{j2}(n_{j2} - R_{j2})} + \frac{n_{j1} - 1}{R_{j1}(n_{j1} - R_{j1})}.$$

If the logistic effect Δ is the same for all k 2×2 tables, the weighted least-squares estimate of Δ is the weighted mean of the separate estimates and is given by (by minimizing $\sum_j (\Delta_j - \Delta)^2/V_{\Delta_j}$ with respect to Δ)

$$\tilde{\Delta}^{(W)} = [\Sigma \tilde{\Delta}_j^{(W)}/V_{\Delta_j}^{(W)}]/[\Sigma(1/V_{\Delta_j}^{(W)})]$$

having the approximate variance $[\Sigma(1/V_{\Delta_j}^{(W)})]^{-1}$. Note that $\tilde{\Delta}^{(W)}$ is not a function of the sufficient statistic for the problem[2]. If the logistic effects Δ_j are the same, then the residual $S_j^{(W)} = [\tilde{\Delta}_j^{(W)} - \tilde{\Delta}^{(W)}]/[V_{\Delta_j}^{(W)}]^{1/2} \approx$ standard normal variable. Thus under H_0,

$$\sum_{j=1}^{k} [\tilde{\Delta}_j^{(W)} - \tilde{\Delta}^{(W)}]^2/V_{\Delta_j}^{(W)} \approx \chi_{k-1}^2.$$

---

[2] If $Y_1, \ldots, Y_n$ are independent Bernoulli random variables with probability of success for Y_i being $P_i = [1 + \exp(-x_i'\beta)]^{-1}$, then the likelihood function can be written as $\exp\left(\sum_{j=1}^{s} \beta_j T_j\right) / \prod_{i=1}^{n} (1 + \exp(x_i'\beta))$ where $T_j = \sum_{i=1}^{n} x_{ij} Y_i$. Thus $(T_1, \ldots, T_s)$ is the sufficient statistic.

Further, the ranked values of $S_j^{(w)}$ can be plotted against the expected standard normal order statistics in a sample of size k.

Other alternative procedures for combining information from several 2×2 tables have been dealt with in chapter 10 of Fleiss [1981]. For point and interval estimation of the common odds ratio from several 2×2 tables, the reader is referred to Gart [1970].

Example 3

Consider the data in Example 2 and some other fictitious data pertaining to the immunity provided to the rats by the antiserum.

Batch	Bacteria and antiserum		Bacteria only	
	total	alive	total	alive
1	57	44	54	29
2	50	38	50	26
3	60	45	60	31
4	40	29	40	21

Logistic analysis of the above data:

Batch	$\bar{\Delta}_j^{(w)}$	$V_{\Delta_j}^{(w)}$	$1/V_{\Delta_j}^{(w)}$	$(1/V_{\Delta_j}^{(w)})^{1/2}$	Residual $S_j^{(w)}$	Logistic sum[1] $= \Sigma Z_j^{(w)}$
1	1.096	0.171	5.848	2.418	0.160	4.64
2	1.043	0.186	5.376	2.319	0.030	4.058
3	1.053	0.156	6.410	2.532	0.058	4.139
4	0.896	0.225	4.444	2.108	−0.282	3.822

[1] Sum of the logistic transforms for bacteria and antiserum, and for bacteria only.

$$\sum_1^4 (1/V_{\Delta_j}^{(w)}) = 22.078,$$

weighted mean $= \bar{\Delta}^{(w)} = 22.748/22.078 = 1.030$,

$$\sum_1^4 S_j^{(w)2} = 0.109 \quad \text{(not significant)}.$$

Also, $\bar{\Delta}^{(w)} (22.078)^{1/2} = 4.84$ (highly significant).

From the preceding analysis, we draw the following conclusions. The data strongly suggest that there are not significant differences between the batches. Thus, they can be combined to obtain a pooled estimate of Δ. When the ranked residuals are plotted against corresponding expected normal order statistics, the points are almost collinear.

The plots of the difference of the logistic transforms against the sum show (see last column) that batch No. 4 not only has a small difference, but also a small sum. We surmise that the effect of the antiserum is highly significant.

Remark. The above data can also be analysed using the likelihood of the observations as functions of the differences $\Delta_1, \ldots, \Delta_k$ (Δ is defined at the beginning of section 6.7). This likelihood approach has been taken by Cornfield [1956] for studying the associations between smoking and lung cancer.

7 Estimation of Points on the Quantal Response Function

Example

Suppose a number of plastic pipes of specified length are subjected to impacts of energies. Several levels of the response may be noted. In particular, we could record whether or not a brittle failure has happened. Let $Y(x)$ denote the response to dose level x and assume that $Y(x)$ takes the value 0 or 1; $P[Y(x) = 1] = P(x)$ where $P(x)$ increases with x. We will be interested in finding x_p for which $P(x_p) = p$. So far we have been dealing with known forms for $P(x)$. Here we assume that $P(x)$ is unknown. This problem is closely related to the Robbins-Monro (R-M) process which is slightly more general and in the following we will review some aspects of the R-M process.

7.1. Robbins-Monro Process

Given a random response $Y(x)$ at x having $EY(x) = M(x)$, we wish to estimate θ such that $M(\theta) = \alpha$.

Robbins and Monro [1951] suggest the following sequential procedure. Guess an initial value x_1 and let $y_r(x_r)$ denote the response at x_r. Then choose x_{n+1} by the formula

$$x_{n+1} = x_n - a_n[y_n(x_n) - \alpha] \quad (n = 1, 2, \ldots) \tag{7.1}$$

where a_r, $r = 1, 2, \ldots$ is a decreasing sequence of positive constants and a_n tends to zero. If we stop after n iterations, then x_{n+1} is the estimate of θ. As a special case, let $\alpha = 0$, then the R-M procedure becomes

$$x_{n+1} = x_n - a_n y_n(x_n). \tag{7.2}$$

Notice that $x_{n+1} < x_n$ if $y_n(x_n) > 0$ and $x_{n+1} > x_n$ if $y_n(x_n) < 0$. An appropriate choice for a_n is c/n, where c is chosen optimally in some sense that will subsequently be made clear. It is not unreasonable to assume that

$M(x) > 0$ for all $x > \theta$,

$M(x) < 0$ for all $x < \theta$.

Derman and Sacks [1959] and Sacks [1958] studied the convergence of x_n, the rate of convergence, the choice of a_n and the asymptotic distribution of x_n. Sacks [1958] has shown that when $a_n = c/n$, the asymptotic distribution of $(x_n - \theta)\sqrt{n}$ is normal $\left(0, \dfrac{\sigma^2 c^2}{2c\beta - 1}\right)$, where $\beta = M'(\theta)$ provided $c\beta > 1/2$.

For a stochastic approximation, we need the following definitions.
Types of convergence: Let $X_1, X_2, \ldots$ be a sequence of random variables.
(1) Mean square convergence. X_n is said to converge to θ in quadratic mean if

$$\lim_{n \to \infty} E(X_n - \theta)^2 = 0$$

(This is also called weak convergence.)

(2) Convergence in probability. For any $\varepsilon > 0$,

$$\lim_{n \to \infty} P(|X_n - \theta| > \varepsilon) = 0.$$

(3) Strong convergence.

$$P\left(\lim_{n \to \infty} X_n = \theta\right) = 1.$$

Note that $(3) \Rightarrow (2)$, $(1) \Rightarrow (2)$.

Definition

Stochastic approximation is concerned with procedures or schemes which converge to some value that is sought, when the observations are subject to error due to the stochastic nature of the problem. Interesting schemes are those that are self-correcting, namely those in which an error always tends to disappear in the limit and in which the convergence to the destined value is of some specified nature (one of the three types mentioned above).

In bioassay, sensitivity testing or in fatigue trials, the statistician is often interested in estimating a given quantile of a distribution function on the basis of some data, each datum being of the 0-1 type.

Example

Let $F(x) = P$ (a metallic specimen will fracture if subjected to x cycles in a fatigue trial).

A specimen tested in such a way will represent an observation and takes either 0 or 1. Then the problem of interest would be to estimate the number of cycles x such that for a given α, $F(x) = \alpha$. Probit analysis and up and down procedures depend upon the functional form of F.

For real x, let $Y(x)$ denote the response to an experiment carried out at a controlled level x and have the unknown distribution function $H(y(x))$ and regression function

$$M(x) = E[Y(x)|x] = \int_{-\infty}^{\infty} y \, dH(y|x).$$

No knowledge of $M(x)$ and $H(y|x)$ is assumed.

7.2. Robbins and Monro Procedure

Let α be a given real number. We wish to estimate the root of the equation, $M(x) = \alpha$, where we assume the existence of a unique root. One can use Newton's method provided the form of the function M is known.

Kiefer-Wolfowitz Process. We wish to find θ such that $M(\theta)$ is a maximum.

Properties of Interest.

(1) To find a recursion for $x_1, \ldots, x_n$ and to show convergence of x_n, that is, to show that $\lim_{n \to \infty} E(x_n - \theta)^2 = 0$,

(2) $P(\lim x_n = \theta) = 1$,

(3) asymptotic normality of x_n or $M(x_n)$,

(4) confidence intervals for θ,

(5) an optimal stopping rule.

Robbins and Monro [1951] proved the following theorems.

Theorem 1

Let $\{a_n\}$ be a sequence of positive constants such that

$$0 < \sum_1^n a_n^2 < \infty \quad \text{and} \tag{1}$$

$$\sum_{n=2}^{\infty} [a_n/(a_1 + \cdots + a_{n-1})] = \infty,$$

which is implied by the condition

$$\sum_1^{\infty} a_n = \infty,$$

$$P(|Y(x)| \leq c) = \int_{-c}^{c} dH(y|x) = 1 \tag{2}$$

for all x, some finite c, and for some $\delta > 0$, $M(x) \leq \alpha - \delta$ for $x < \theta$ and $M(x) \geq \alpha + \delta$ for $x > \theta$. Then the sequence

$$x_{n+1} = x_n + a_n(\alpha - y_n)$$

converges in quadratic mean and hence in probability, i.e.

$\lim_{n \to \infty} b_n = \lim_{n \to \infty} E(x_n - \theta)^2 = 0$ and $x_n \to \theta$ in probability.

Theorem 2 (Special Case of Theorem 1)

If (1) and (2) of Theorem 1 hold and $M(x)$ is nondecreasing, $M(\theta) = \alpha$, $M'(\theta) > 0$, then $\lim_{n \to \infty} b_n = 0$.

Example

Let F(x) be a distribution function such that

$F(\theta) = \alpha \ (0 < \alpha < 1), F'(\theta) > 0,$

and $\{Z_n\}$ be a sequence of independent random variables such that $P(Z_n \leq x) = F(x)$. We are not allowed to know the values of Z_n, but for each n we are free to specify a value x_n and we observe only the values $\{y_n\}$ where

$$y_n = \begin{cases} 1 & \text{if } Z_n \leq x_n \quad \text{'response'} \\ 0 & \text{if } Z_n > x_n \quad \text{'no response'}. \end{cases}$$

Here $M(x) = F(x)$ and other conditions are satisfied. Robbins and Monro [1951] feel that the assumption of boundedness of Y(x) with probability 1 for all x is somewhat too strong. Wolfowitz [1952] has shown the convergence of x_n to θ in probability under weaker conditions. Blum [1954] provided the least restrictive conditions for the convergence of x_n to θ with probability 1.

7.2.1. Blum's Conditions

$|M(x)| \leq c + d|x|$ for some $c, d \geq 0$ (1)

$\sigma_n^2 = \int_{-\infty}^{\infty} [y - M(x)]^2 \, dH(y|x) \leq \sigma^2 < \infty$ (2)

$M(x) < \alpha$ when $x < \theta$ (3)

$M(x) > \alpha$ when $x > \theta$.

$\underset{\delta_1 \leq (x-\theta) \leq \delta_2}{\text{Inf}} [M(x) - \alpha] > 0$ for every $\delta_1, \delta_2 > 0$. (4)

Under the above assumptions Blum [1954] showed that

$$P\left(\lim_{n \to \infty} x_n = \theta\right) = 1.$$

Notice that assumption (4) allows the possibility of $M(x) \to \alpha$ as $|x| \to \infty$ and in such a case one would expect that there might be a positive probability of $|x_n|$ converging to ∞. Dvoretzky [1956] has shown that under the conditions of Blum [1954] x_n converges in the mean to θ. Derman [1956] points out that using Kolmogorov's inequality one can show that (since $\Sigma a_n = \infty$, $\Sigma a_n^2 < \infty$ and $\sigma_x^2 \leq \sigma^2 < \infty$)

$$\sum_{j=1}^{\infty} a_j [y_j - M(x_j)]$$

converges with probability 1 and consequently

$$x_{n+1} - \sum_{j=1}^{n} a_j[\alpha - M(x_j)] = x_1 - \sum_{j=1}^{n} a_j[y_j - M(x_j)]$$

$$\left(\text{since } \sum_{j=1}^{n} a_j(y_j - \alpha) = \sum_{j=1}^{n}(x_j - x_{j+1}) = x_1 - x_{n+1}\right)$$

converges with probability 1.

7.2.2. Asymptotic Normality

Chung [1954], Hodges and Lehmann [1956], and Sacks [1958] have considered the asymptotic normality of x_n. Under certain regularity assumptions, Chung [1954] has shown that $n^{1/2}(x_n - \theta)$ is asymptotically normal $[0, \sigma^2 c^2/(2\alpha_1 c - 1)]$ when $na_n \to c$ and where $\alpha_1 = M'(\theta) > 0$, $c > (2K)^{-1}$ and $K \leq \inf[M(x) - \alpha]/(x - \theta)$. Hodges and Lehmann [1956] recommend taking $c = 1/\alpha_1$ since it minimizes $\sigma^2 c^2/(2\alpha_1 c - 1)$, the minimum value being σ^2/α_1^2. (In practice one has to guess the value of α_1.) Since $K \leq \inf[M(x) - \alpha]/(x - \theta)$ where the infinium can be restricted to the value $|x - \theta| \leq A$, where A is an arbitrarily small number and since $M'(\theta) = \alpha_1 > 0$, it suffices to require that $c > 1/(2\alpha_1)$. Hodges and Lehmann [1956] also show that the condition $\lim M(x)/x > 0$ as $|x| \to \infty$ is not necessary. Also, in practice we might be tempted to use a 'safe' small prior estimate for α_1 and hence a correspondingly large c in order to avoid the possibility of $c \leq 1/(2\alpha_1)$. In this case the estimates would have unknown behavior. In the following we shall use the alternative approach of Hodges and Lehmann [1956] which works for all values of $c > 0$ and provides measures of precision for finite n.

7.2.3. Linear Approximation

Here the actual model is replaced by a linear model for which the exact error variances can be evaluated. We assume that $M(x) = \alpha + \beta(x - \theta)$ and $V(x) = \text{var}(Y|x) = \tau^2$, where β and τ are known constants. We might take $\beta = \alpha_1$ and $\tau^2 = \sigma^2 [= V(\theta)]$. The justification for such an approximation is that $M(x)$ is linear in the vicinity where the x_n is likely to fall. Also without loss of generality we can set $\alpha = \theta = 0$. That is, $M(x) = \beta x$ and $\text{var}[Y(x)] \equiv \sigma^2$. Since

$$x_{n+1} = x_{n+1}(x_n, y_n) = x_n - a_n y_n(x_n),$$

we have

$$E(x_{n+1}|x_n) = (1 - a_n\beta)x_n,$$

and hence by iteration

$$E(x_{n+1}) = x_1 \prod_{r=1}^{n}(1 - a_r\beta),$$

where x_1 denotes the initial value. The bias in x_{n+1} could be zero if $x_1 = 0$. Also squaring x_{n+1} and taking conditional expectations, we obtain

$$E(x_{n+1}^2 | x_n) = (1 - \beta a_n)^2 x_n^2 + a_n^2 \sigma^2$$
$$= (1 - \beta a_n)^2 (1 - \beta a_{n-1})^2 x_{n-1}^2 + \sigma^2 [a_n^2 + a_{n-1}^2 (1 - \beta a_n)^2].$$

Hence

$$E(x_{n+1}^2) = x_1^2 \left[\prod_{r=1}^{n} (1 - \beta a_r) \right]^2 + \sigma^2 \sum_{r=1}^{n} a_r^2 \prod_{s=r+1}^{n} (1 - \beta a_s)^2.$$

Now, set $a_n = c/n$ and obtain

$$E(x_{n+1}) = x_1 \phi_n(c\beta), \text{ and}$$

$$E(x_{n+1}^2) = x_1^2 \phi_n^2(c\beta) + \frac{\sigma^2}{\beta^2} \psi_n(c\beta) \quad \text{where}$$

$$\phi_n(z) = (n!)^{-1} \prod_{1}^{n} (r - z) = \Gamma(n + 1 - z)/\Gamma(1 - z)n! \quad \text{and}$$

$$\psi_n(z) = \sum_{r=1}^{n} \frac{z^2}{r^2} \prod_{s=r+1}^{n} (1 - z/s)^2 = \frac{\Gamma^2(n + 1 - z)}{(n!)^2} \sum_{r=1}^{n} \frac{z^2 (r!)^2}{r^2 \Gamma^2 (1 + r - z)}.$$

Wetherill [1966, p. 148] has provided more details for getting asymptotic approximations to $\phi_n(z)$ and $\psi_n(z)$ which will be given below. The first term in the expression for $E(x_{n+1}^2)$ is due to the bias in x_{n+1} which vanishes when $c\beta$ is an integer and $c\beta \leq n$. Asymptotically

$$\phi_n(z) = (1 - z/n)^{1/2} / n^z \Gamma(1 - z).$$

Thus, the contribution of the bias term to the mean squared error of x_{n+1} is of the order $0(n^{-2c\beta})$. The evaluation of the second term $\psi_n(t)$ is somewhat complicated. Applying Stirling's formula, we have

$$\left[\frac{\Gamma(n + 1 - z)}{\Gamma(n + 1)} \right]^2 \doteq (1 - z/n)(n - z)^{-2z} = [n(n - z)^{2z-1}]^{-1}.$$

which can be used in the summation part of $\psi_n(z)$. Let

$$S(z) = \sum_{r=1}^{n} r^{-2} \left[\frac{\Gamma(r + 1)}{\Gamma(r + 1 - z)} \right]^2 = \sum_{r=1}^{n} r^{-2} r(r - z)^{2z-1} = \sum_{r=1}^{n} (r - z)^{2z-1}/r.$$

If $z = \frac{1}{2}$, the last summation is asymptotically equivalent to $\ln n$. For $z > \frac{1}{2}$

$$S(z) \doteq \int_{1}^{n} \frac{(r - z)^{2z-1}}{r} dr \approx \int_{1}^{n} [(r - z)^{2z-2} - z(r - z)^{2z-3} + \cdots] dr$$

$$\approx (n - z)^{2z-1}/(2z - 1). \quad \text{Thus}$$

$$E(x_{n+1}^2) \doteq \begin{cases} \sigma^2 \log_e n / (4\beta^2 n) & \text{if } c\beta = \frac{1}{2} \\ \sigma^2 c^2 / n(2c\beta - 1) & \text{if } c\beta > \frac{1}{2}. \end{cases}$$

Thus, when $c\beta > \frac{1}{2}$, the mean squared error of x_n is $0(n^{-1})$ and it coincides with the asymptotic variance obtained by Chung [1954] and Sacks [1958] for the quasi-linear case, namely $M(x) = \alpha + \beta(x - \theta) + o(|\beta - \theta|)$. Sacks [1958] proved the asymptotic normality of $(x_n - \theta)$, under very general conditions, having mean zero and variance $\sigma^2 c^2/n(2c\beta - 1)$ with $a_n = c/n$ and $\beta = M'(\theta)$ when $2\beta c > 1$. For $c = 0.2\ (0.2)\ 0.8$ and 1.2, $n = 5\ (5)\ 30$; $n\phi_n(c)$ are tabulated by Hodges and Lehmann [1956] and for $c = 0.2\ (0.2)\ 0.8$, $1.2\ (0.4)\ 2.0$ and 3.0, $n = 5\ (5)\ 30$, ∞, the value of $n\psi_n(c)$ is also tabulated. Also note that $n\phi_n(1) = 0$, while $n\phi_n^2(\frac{1}{2}) \to 1/\pi = 0.318$ due to Wallis' product. For $n \geq 30$, one can use

$$\phi_n(z) = \phi_{30}(z)(30.5n^{-1} + 2^{-1})^{2z}.$$

Also we have the recursion formula

$$\psi_n(z) = (z/n)^2 + [(1-z)/n]^2 \psi_{n-1}(z).$$

Using a quadratic interpolation formula on $n\psi_n(z)$ against $1/n$ at the values $1/n = 0, \frac{1}{2}, 1$ one obtains

$$n^3 \psi_n(z)/z^2 = (n-1)(n-2)(2z-1)^{-1} + 2(2-z)^2(n-1) + n.$$

The shortcoming of the linear model is that we do not know how approximately linear $M(x)$ is and how nearly constant $V(x)$ is.

7.3. *Parametric Estimation*

Although the R-M procedure is completely nonparametric since it does not assume any form for $M(x)$ or for $H(y|x)$, in several cases, especially in quantal response situations, $H(y|x)$ is known (is Bernoulli) except for the value of a real parameter γ. We can reparametrize in such a way that γ is the parameter to be estimated. Let $E[Y(x)] = M_\gamma(x)$, $V[Y(x)] = V_\gamma(x)$ and let these functions satisfy the conditions imposed earlier. Since γ determines the model, θ is a function of α and γ. We further assume that there is a 1-1 correspondence between θ and γ so that there exists a function h such that $\gamma = h_\alpha(\theta)$. Then we may use x_n as the R-M estimate of θ and obtain the estimate of γ as $h_\alpha(x_n)$. Now the problem is choosing a_n (and perhaps α also) in order to minimize the asymptotic normal variance of the estimate of γ. One can easily show that $n^{1/2}[h_\alpha(x_n) - \gamma]$ is asymptotically normal with mean 0 and variance $[h'_\alpha(\theta)]^2 \sigma^2 c^2/(2\alpha_1 c - 1)$.

Example

Consider the quantal response problem in which $Y(x) = 0$ or 1 with $P[Y(x) = 1] = M_\gamma(x)$ and $V_\gamma(x) = M_\gamma(x)[1 - M_\gamma(x)]$. By choosing $0 < \alpha < 1$ we estimate θ by means of the R-M procedure, which yields a sequence of estimates x_n such that $n^{1/2}(x_n - \theta) \approx$ normal $(0, [V_\gamma(\theta)c^2/(2\alpha_1 c - 1)])$. Let

$\partial M_\gamma(x)/\partial x = M'_\gamma(x), \quad \partial M_\gamma/\partial \gamma = M^*_\gamma(x).$

For given α, the best value of $c = \lim na_n$ is $[M'_\gamma(\theta)]^{-1}$. With this c, the asymptotic variance of $n^{1/2} h_\alpha(x_n)$ is

$[h'_\alpha(\theta)]^2 \sigma^2(\theta)/[M'_\gamma(\theta)]^2 = \alpha(1 - \alpha)/[M^*_\gamma(\theta)]^2$

since $\sigma^2(\theta) = V_\gamma(\theta) = \alpha(1 - \alpha)$ and by differentiating the identity $M_{h_\alpha(\theta)}(\theta) = \alpha$ w.r.t. θ, we obtain $h'_\alpha(\theta) = -M'_\gamma(\theta)/M^*_\gamma(\theta)$. Now the value of α, which minimizes $\alpha(1 - \alpha)/[M^*_\gamma(\theta)]^2$, is independent of γ provided that $M^*_\gamma(\theta)$ factors into a function of θ [like $M_\gamma(x) = r(s(\gamma))t(x)$]. For example, we can take

$P[Y(x) = 1] = M_\gamma(x) = F[x - \gamma + F^{-1}(\beta)] \quad \text{for some } 0 < \beta < 1,$

where F is a distribution function. Now γ can be interpreted as the dose x for which the probability of response is β. That is, $\gamma = LD_{100\beta}$ (lethal dose 100 β). Then the formula for the asymptotic variance becomes

$\alpha(1 - \alpha)/\{F'[F^{-1}(\alpha)]\}^2$

since $M_\gamma(\theta) = \alpha$ implies $F[\theta - \gamma + F^{-1}(\beta)] = \alpha$ which implies $F^{-1}(\alpha) = \theta - \gamma + F^{-1}(\beta)$, and $M^*_\gamma(\theta) = -F'[\theta - \gamma + F^{-1}(\beta)]$. The asymptotic variance is independent of β since the problem is invariant under location shifts. Now the value of α that minimizes the asymptotic variance is $\alpha = \frac{1}{2}$ when F is normal or logistic.

Suppose we want to estimate $\gamma = LD_{100\beta}$. Then we do not need the parametric model since we may set $\alpha = \beta$ and $\gamma = \theta$ and thus estimate γ directly from x_n via the R-M scheme. The advantage of this method is that it assumes very little about the form of F, the disadvantage may be a significant loss of efficiency, especially when β is not near $\frac{1}{2}$.

Example

Suppose we wish to estimate the mean bacterial density γ of a liquid by the dilution method. For a volume x of the liquid, let $Y(x) = 1$ if the number of bacteria is 1 or more. Then $P[Y(x) = 1] = M_\gamma(x) = 1 - \exp(-\gamma x)$ under the Poisson model. Then $M^*_\gamma(\theta) = \theta \exp(-\gamma \theta) = -(1 - \alpha)\ln(1 - \alpha)/\gamma$ since $1 - \exp(-\gamma \theta) = \alpha$. Thus, the asymptotic variance becomes $[\alpha/(1 - \alpha)\log^2(1 - \alpha)]\gamma^2$ and whatever be γ, this is minimized by minimizing the first factor. The best α is the solution of the equation $2\alpha = -\log(1 - \alpha)$ or $\alpha = 0.797$. Thus, the recommended procedure is: carry out the R-M scheme with $\alpha = 0.797$ and $a_n = 4.93/\hat{\gamma}n$ [since $1/c = \alpha_1 = M'_\gamma(\theta) = \gamma(1 - \alpha)$] where $\hat{\gamma}$ is our prior estimate of γ. Our estimate for γ after n steps is $2\alpha/x_{n+1} = 1.594/x_{n+1}$ [since $\gamma = h_\alpha(\theta) =$

$-\ln(1-\alpha)/\theta]$ and the asymptotic variance is $\gamma^2/4\alpha(1-\alpha) = 1.544 \gamma^2$ since $2\alpha = -\ln(1-\alpha)$.

Block [1957] considers replication at each x_i and replacing y_i by $\bar{y}_{n_i}$ when n_i is the subsample size at x_i. The rationale behind this is that it may be cheaper to observe a set of observations at a dose level x_i rather than the same number of observations at different dose levels.

Maximum Likelihood Estimate of θ

Let $f(x_1,\ldots,x_n)$ denote the joint density of $(x_1,\ldots,x_n)$. If $y_n = M(x_n) + \varepsilon_n = \alpha + \alpha_1(x_n - \theta) + \varepsilon_n$ and if the ε_i are normal $(0,\sigma^2)$, then the joint density function of $x_1, \ldots, x_n$ and $y_1, \ldots, y_n$ is

$$(\sqrt{2\pi}\sigma)^{-n} \exp\left[-\frac{1}{2\sigma^2}\Sigma(y_i - \alpha - \alpha_1(x_i - \theta))^2\right] f(x_1,\ldots,x_n)$$

where σ^2 is known and θ is to be estimated. The mle of θ is $\hat{\theta} = \bar{x} - \alpha_1^{-1}(\bar{y} - \alpha)$ which is distributed normally $(\theta, \sigma^2/\alpha_1^2 n)$ (first condition on $x_1, \ldots, x_n$ and then uncondition). Thus, the R-M estimator with the optimal choice of c is asymptotically efficient.

Venter [1967] proposed a modification of the Robbins-Monro (R-M) procedure of which the asymptotic variance is the minimum even when $M'(\theta)$ is unknown. He replaces the constant c in the R-M procedure by a sequence which converges to $[M'(\theta)]^{-1}$. Although Venter's [1967] modification requires two observations at each stage, the asymptotic variance of Venter's modification based on n (total) observations with $M'(\theta)$ unknown is the same as that of the R-M procedure under Sack's [1958] assumptions with n observations with $M'(\theta)$ known. Rizzardi [1985] has shown that the original and Venter's modification of R-M estimator are locally asymptotically minimax, for a class of regression functions that includes distribution functions in the case where the response is a 0-1 variable. Rizzardi [1985] studied some stochastic approximation procedures for estimating the ED_p, namely, the effective dosage level to obtain a preassigned probability p of response when the response curve is assumed to be logistic and the response is a 0-1 variable. He compares the estimators by their expected squared error loss. In particular, when the tolerence distribution is logistic, and the parameter of interest is ED_{50}, namely the median of the distribution, Rizzardi [1985], based on numerical computations with sample sizes n = 10, 20, infers that the expected squared error loss of the R-M procedure and Venter's modification are very close to the asymptotic variance. He also studied the problem of estimating the difference of the medians of two logistic distributions when the scale parameters are (1) equal and unknown and (2) unequal and unknown.

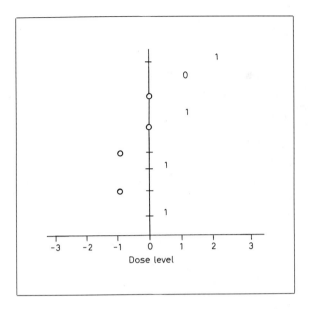

Fig. 2

7.4. Up and Down Rule

Dixon and Mood [1948] have proposed a method of estimating LD_{50} which is simpler than the R-M procedure. In the latter, the dose levels are random and we cannot specify them ahead of time. The Dixon-Mood procedure chooses a series of equally spaced dose levels x_i. $Y(x)$ is observed only at these levels. We take the first observation at the best initial guess of LD_{50}. If the response is positive, the next observation is made at the immediately preceding lower dose level. If the response is zero, the next trial is made at the immediately proceeding dose level. If the positive response is coded as 1 and no response is denoted by 0, then the data may look like the one in figure 2. (Suppose we start at 0, let the response be 1; then we go to dose -1 and the response is 0; then go to dose 0, and the response is 1; then go to dose -1 at which the response is 0; then go to zero dose at which the response is 0; then go to dose 1 and suppose the response is 1; then go to dose 0, the response is 0; then take dose 1, the response is 0; then go to dose 2, the response is 1 etc.)

The main advantage of this up and down method is that it automatically concentrates testing near the mean. Furthermore, this increases the accuracy of the estimate. The savings in the number of observations may be 30–40%. Another advantage is that the statistical analysis here is quite simple. One possible disadvantage is that it requires each specimen be tested separately, which is not economical, for instance, in tests of insecticides.

The method is not appropriate for estimating dose levels other than LD_{50} because of the assumption that the normal or the logistic distribution will be violated in the extreme tails of the distribution. Also, the sample size is assumed to be large since only large-sample properties of the estimator will be asserted. Also, one must have a rough estimate of the standard deviation in advance, because the intervals between testing levels should be equal to the standard deviation. This condition will be satisfied if the interval actually used is less than twice the standard deviation.

If y denotes the response, let y_0 be an estimate of LD_{50}. Then the tests are made at $y_i = y_0 \pm id$ $(i = 0, 1, 2, \ldots)$ where d is an estimate of the standard deviation and y_0 is the level of the initial test. Let n_i denote the number of successes and m_i the number of failures at y_i $(i = 0, 1, \ldots)$. Then the distribution of these variables is

$$p(n, m | y_0) = K \prod_{i=-\infty}^{+\infty} p_i^{n_i} q_i^{m_i}, \quad n = (n_1, n_2, \ldots), \quad m = (m_1, m_2, \ldots), \quad \text{where}$$

$p_i = \Phi[(y_i - \mu)/\sigma] = 1 - q_i$ $(i = 0, 1, \ldots)$ for the probit case,

$ = \{1 + \exp[-(\alpha + \beta y_i)]\}$ for the logit case,

and K does not involve the unknown parameters. The method of maximum likelihood will be used in order to estimate the unknown parameters. Since $|n_i - m_{i-1}| = 0$ or 1 for all i, either one of the sets $\{n_i\}$ or $\{m_i\}$ contains practically all the information in the sample. If $N = \Sigma n_i$ and $M = \Sigma m_i$, and assuming that $N \leqslant M$, we can write the approximate likelihood function as

$$p(m, n | y_0, M - N) = K' \prod_i (p_i q_{i-1})^{n_i}.$$

Dixon and Mood [1948] maximize the above with respect to the unknown parameters μ and σ and obtain the estimates.

Let $P_i(k)$ denote the probability that the k-th observation is taken at y_i. If the k-th observation is at y_i, then the $(k-1)$-th observation was at either y_{i+1} or y_{i-1}. Hence we have

$$P_i(k) = P_{i+1}(k-1)p_{i+1} + P_{i-1}(k-1)q_{i-1}$$

with boundary conditions $P_0(1) = 1$, $P_i(1) = 0$ for $i \neq 0$. Also, because $|n_i - m_{i-1}| = 0$ or 1, asymptotically the P_i satisfy (by writing $p_i = P_i p_i + P_i q_i$ and taking $P_{i+1} \doteq P_i$ and $p_{i+1} \doteq q_i$), $P_i p_i = P_{i-1} q_{i-1}$. Hence

$$P_i = P_0 \prod_{j=0}^{i-1} q_j \bigg/ \prod_{j=1}^{i} p_j, \quad \text{for } i > 0$$

$$ = P_0 \prod_{j=i+1}^{0} p_j \bigg/ \prod_{j=i}^{-1} q_j, \quad \text{for } i < 0.$$

Then $E(n_i)$ can be obtained from the relation $E(n_{i+1})/q_i = E(n_i)/p_i$. Hence it follows that

$$E(n_i) = NW_i \bigg/ \sum_{-\infty}^{\infty} W_i, \quad \text{where } W_i = \prod_{j=0}^{i-1}(q_j/p_j), \quad \text{for } i > 0, \text{ and}$$

$$W_i = \prod_{j=-1}^{i}(p_j/q_j), \quad \text{for } i < 0.$$

7.4.1. mle Estimates of μ and σ

For estimation purposes, only the successes or only the failures will be used depending on which has the smaller total. Assume that $N < M$. Then, let

$A = \Sigma i n_i, \quad B = \Sigma i^2 n_i.$

Then, Dixon and Mood [1948] obtain

$m = \hat{\mu} = y' + d[(A/N) \pm (1/2)]$

where y' is the normalized level corresponding to the lowest level at which the less frequent event occurs. The plus sign (minus sign) is used when the analysis is based on the failures (successes). Also,

$s = \hat{\sigma} = 1.620d[(NB - A^2)N^{-2} + 0.029],$

which is quite accurate when $(NB - A^2)/N^2$ is larger than 0.3. When $(NB - A^2)N^{-2}$ is less than 0.3, Dixon and Mood [1948, appendix B] give an estimate of σ based on an elaborate calculation. They also provide confidence intervals for μ that are based on the large sample properties of the mle values.
Brownlee et al. [1953] obtain an estimate which is asymptotically equivalent to the approximation of Dixon and Mood [1948] and to the maximum likelihood estimator and is given by

$$\hat{L}_{0.50} = n^{-1} \sum_{r=2}^{n+1} y_r$$

for a sequence of n trials, where y_r is the level used at the r-th trial. (Notice that the level of the first trial gives no information, although the level at which we would have taken the $(n + 1)$-th observation certainly has some information.)

7.4.2. Logit Model

For the logit model, one can obtain the mle of β and $-\alpha/\beta$ and their standard errors based on the method of maximum likelihood. For some new alternative strategies and studies via Monte Carlo trials, the reader is referred to Wetherill [1966, Section 10.3].

7.4.3. Dixon's Modified Up and Down Method

The up and down procedure proceeds in the manner described earlier until the nominal sample size is reached. The nominal sample size N^* of an up and down sequence of trials is a count of the number of trials, beginning with the first pair of responses that are unalike. For example, in the sequence of trial result 000101, the nominal sample size is 4. Although the similar responses preceding the first pair of changed responses do not influence N^*, they are used in estimating the ED_{50}. For $1 < N^* \leq 6$, the estimate of ED_{50} is obtained from table 2 of Dixon [1970, p. 253] and is given by

$$\log ED_{50} = y_f + kd$$

where y_f denotes the final dose level in an up and down sequence and k is read from table 2 of Dixon [1970]. For example, if the series is 011010 and $y_f = 0.6$ and $d = 0.3$, then $N^* = 6$ and the estimate of $\log ED_{50}$ is $0.6 + 0.831 (0.3) = 0.85$. For nominal sample sizes greater than 6, the estimate of $\log ED_{50}$ is

$$(\Sigma y_i + dA^*)/N^*$$

where the y_i values are the log dose levels among the N^* nominal sample size trials, and where A^* is obtained from table 3 of Dixon [1970, p. 254] and A^* depends on the number of initial-like responses and on the difference in the cumulative number of 1 and 0 values in the nominal sample size N^*.

In order to plan an up and down experiment, one needs to specify (1) starting log dose, (2) the log dose spacing, and (3) the nominal sample size. It is desirable to have the starting dose as close to the ED_{50} as possible, because further the starting dose is from the ED_{50}, the greater is the likelihood of ending up with a string of similar responses prior to beginning of the nominal sample size. The choices for equal log dose spacings are $2\sigma/3$, σ, $3\sigma/2$. Also, N^* should be 3, 4, 5 or 6, because the mean square error of the up and down estimate of ED_{50} for $3 < N^* \leq 6$ is essentially independent of the starting dose and is approximately $2\sigma^2/N^*$, where σ^2 denotes the variance of the underlying population.

An additional factor which is in favor of the up and down method is that its analysis procedure minimizes the sum of squares on log dose rather than on response. The computations are relatively easy and the interpretation follows the well-known regression analysis.

The asymptotic maximum likelihood solution given by Dixon and Mood [1948] did not make full use of the initial scores of trials and the occurrence of an unequal number of 0 and 1. The estimate of Brownlee et al. [1953] gave a reduction in MSE by adjusting the sample size for series starting at some distance from the true ED_{50}. However, both the above estimates do depend on the initial dose level.

8 Sequential Up and Down Methods

8.1. *Up and Down Transformed Response Rule*

Wetherill [1963] and Wetherill et al. [1966] proposed an up and down transformed response rule (UDTR) for estimating LD (100p), $p > \frac{1}{2}$.

Use a fixed series of equally spaced dose levels and after each trial estimate the proportion p' of positive responses (which are denoted by 1) at the dose level used for the current trial, counting only the trials after the last change of the dose level. If $p' > p$, and p' is estimated from n_0 trials or more, go to the immediately lower dose level. If $p' < p$, go to the immediately higher dose level, and if $p' = p$, do not change the dose level. If $p < \frac{1}{2}$, UDTR is modified accordingly by calculating the proportion of negative rather than positive responses. For instance, if $n_0 = 3$, and $p > 0.67$, define the responses as type U or type D:

U = (0), (1,0), (1,1,0); D: (1,1,1).

The first trial is performed at an arbitrary level x and more trials are carried out at the same level, if necessary, until we have a type U or type D response, and then move one level up or down respectively. If the consecutive results at any level are simply classified as type U or D, the following set of UDTR results can be classified as follows.

(1) Results in original form ending in 2nd change:

Level	Observation No.											
	1	2	3	4	5	6	7	8	9	10	11	12
3												
2	1	1	1							1	1	1
1				1	1	1		1	0			
0							0					

(2) Results classified as U or D ending at 6th change:

Level	Change No.					
	1	2	3	4	5	6
3	—	—	—	—	—	D
2	D	—	—	D	—	U
1	—	D	U	—	U	—
0	—	—	U	—	—	—

If the probability of a positive response at any level j is $P(x_j)$ then the probability that the trial at this level will result in a change of level downward is $[P(x_1)]^3$. If responses are classified as U or D, then the response curve is

$F(x) = [P(x)]^3$.

8.2. Stopping Rules

Sometimes we are restricted to take only a fixed number of observations. However, often some flexibility is allowed as regards the number of observations, in order to compensate for poor initial guesses by taking additional observations. Hence, it is desirable to have some stopping rules. Wetherill [1966] give two stopping rules.
Rule 1. Stop after a given number of changes of response type.
Rule 2. Stop when the maximum overdose levels of the number of trials per level reaches a specified number.
Rule 2 may not be easy to apply, therefore rule 1 is recommended.

8.3. Up and Down Method

Alternative Procedures for Estimating LD_{50}. Let $x_0, x_1, \ldots$ denote the dose levels and we sample according to the up and down method of Dixon and Mood [1948] until k (k ⩽ n) turning points (peaks and valleys) are observed. For instance, consider figure 3 [Choi, 1971], where o = negative response and ● = positive response. The method of estimating LD_{50} of a response curve, based on data depicted in figure 3 has been considered by Dixon and Mood [1948], Brownlee et al. [1953], Wetherill [1963], Wetherill et al. [1966], Dixon [1965], Cochran and Davis [1964], Tsutakawa [1967], Hsi [1969], and Choi [1971]. Recall that Brownlee et al. [1953] propose

$$\hat{r} = (n + 1)^{-1} \sum_{j=1}^{n+1} x_j$$

where x_{n+1} is the dosage that would be used at the (n + 2)-th trial if the experiments were continued. Wetherill et al. [1966] propose $\bar{w}$ and $\hat{w}$ that are based only on the peaks and valleys of the response series. Let $\bar{x}_i$ denote the halfway value between the i-th turning point and the dosage level used at the immediately preceding trial (1 = 1, 2, ..., k). Then

$$\bar{w} = \sum_{i=1}^{k} \bar{x}_i/k.$$

$\hat{w}$ is defined as the average of the dosages based at the turning points themselves. For instance, the data in figure 3 yield k = 7, δ(spacing) = 1.

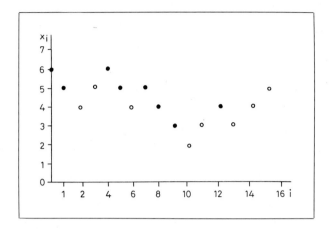

Fig. 3

$\bar{x}_1 = 4.5, \bar{x}_2 = 5.5$ etc. and

$\bar{w} = 4.07, \hat{w} = (4 + 6 + \cdots + 3)/7 = 4.0$

while $\hat{\mu} = 4.07$. One can easily see that

$\hat{w} = \bar{w}$ if k is even

$\quad = \bar{w} \pm \delta/2k$ if k is odd,

because, when k is even, the number of peaks is equal to the number of valleys; whereas, when k is odd, the number of peaks differs at most by 1 from the number of valleys. Using the Markov chain model approach discussed earlier by Wetherill et al. [1966] to the set of turning points, Choi [1971] was able to obtain the variances of $\bar{w}$ and $\hat{w}$.

The phasing factor defined as the distance of μ to the closest dosage level plays an important role. Also, the question of spacing for the estimates $\bar{w}$ and $\hat{w}$ is of practical interest, since the smaller the spacing, the larger is the average sample number (ASN) required to yield the specified number of turning points. Dixon [1965] suggests to use ASN times the mean square error (MSE) of an estimator as a criterion. Based on several exact and Monte Carlo studies Choi [1971] reaches the following conclusions.

(1) Like $\hat{\mu}$, the estimators $\bar{w}$ and $\hat{w}$ are biased although the bias decreases as k increases.

(2) It is more advantageous to obtain an odd number of turning points before termination of the experiment.

(3) On the basis of Dixon's [1965] criteria, this optimal spacing seems to be $\delta = 0.8\sigma$ for any k (σ^2 denoting the variance of the population).

(4) $x_0 > \mu$ if and only if $x_0 > E(\bar{w})$ and $x_0 - E(\bar{w})$ is a monotone function of $x_0 - \mu$. The same property holds for $\hat{w}$.

(5) $x_0 > E(\bar{w})$ if and only if $E(\bar{w}) > \mu$, and $E(\bar{w}) - \mu$ is a monotone function of $x_0 - E(\bar{w})$. Analogous property holds for $\hat{w}$.
(6) $\text{Var}\,\bar{w} \leq \text{var}\,\hat{w}$, however bias $(\bar{w}) \geq \text{bias}(\hat{w})$.
(7) Both $\bar{w}$ and $\hat{w}$ appear to be somewhat more efficient than $\hat{\mu}$ on the basis of MSE if $\delta = 0.5\sigma$, but not when $\delta = 1.0\sigma$. However, uniform superiority of any one of the three estimators cannot be established.

8.4. Finite Markov Chain Approach

Wetherill et al. [1966] point out that the estimation by $\bar{w}$ or $\hat{w}$ can be viewed in terms of a Markov chain. Define the state of a sequence of observations only when there has been a change of response type. Let us abbreviate the state which is a peak by p and a valley by v. Usually there will be an infinite number of states, however, we can impose the restrictions such as $P(z) = 1.0$ for $z \geq 10$ and $P(z) = 0$ for $z \leq -10$ and assume there are a finite number m of possible states and the systems of states will be denoted by

$$\{S_1^p, S_2^p, \ldots, S_m^p, S_1^v, \ldots, S_m^v\}.$$

Let

P (positive response at dose level i) = P_j and

P (negative response at level j) = Q_j (j = 1,...,m).

Also, without loss of generality let the levels be numbered in the decreasing order so that $P_1 > P_2 > \ldots > P_m$ and because of the preceding restriction, $P_m = 0$ and $P_1 = 1$.

If the sequence of observations is at a state S_i^p, then only valleys $(S_{i+1}^v, \ldots, S_m^v)$ are possible for the next state. Hence

$$t_{ij} = P(\text{arriving at } S_j^v \text{ from } S_i^p) = Q_j \prod_{r=i+1}^{j-1} P_r, \text{ for } i < j, \text{ and zero elsewhere, and}$$

$$\tau_{ij} = P(\text{arriving at } S_j^p \text{ from } S_i^v) = P_j \prod_{r=j+1}^{i-1} Q_r, \text{ for } i > j, \text{ and zero elsewhere.}$$

Thus, the matrix of transition probabilities of the chain is

$$T = \begin{bmatrix} 0 & : & B \\ \cdots & & \cdots \\ A & : & 0 \end{bmatrix} \quad \text{where}$$

$$A = \underset{m \times m}{(t_{ij})}, \quad B = \underset{m \times m}{(\tau_{ij})}$$

and A is the lower triangular and B the upper triangular.

The Markov chain is periodic with period 2. There is strong dependence between successive x_i and between averages of successive pairs of x_i.

For a sequence started at any given state, let

$$a^{(n)} = [a_1^{(n)}, \ldots, a_m^{(n)}], b^{(n)} = (b_1^{(n)}, \ldots, b_m^{(n)})$$

denote the probabilities that the n-th turning point is a specified peak or valley, respectively. Obviously $a_m^{(n)} = 0$, since there is no peak at level m and $b_1^{(n)} = 0$ since there is no valley at level 1 (n = 1, 2, ...). One can easily establish the following recurrence relations for a and b

$$a^{(n+1)} = Bb^{(n)} \quad \text{and} \quad b^{(n+1)} = Aa^{(n)} \quad \text{or}$$

$$a^{(n+2)} = BAa^{(n)} \quad \text{and} \quad b^{(n+2)} = ABb^{(n)}.$$

As the sample size increases, $a^{(n)}$ and $b^{(n)}$ approach a steady state and the asymptotic distribution of peaks and valleys can be obtained by computing the principal eigen vectors of BA and AB. Also, one can obtain the exact small sample distribution of peaks and valleys and of the estimators $\bar{w}$ and $\hat{w}$. Wetherill et al. [1966] calculate the mean of $\bar{w}$ and Choi [1971] computes the variance of $\bar{w}$. Notice that the UDTR of Wetherill [1963] can be regarded as an up and down procedure if performed on a transformed response curve. Thus, the method of estimation and Markov chain theory etc. carry over to UDTR.

8.5. Estimation of the Slope

Let $P_i = F(\alpha + \beta x_i)$.

Suppose we are interested in estimating $\tau = 1/\beta$. We conduct two separate experiments; one to estimate LD (100p) and the other to estimate LD (100q) where $q = 1 - p$, $p > \frac{1}{2}$. In both cases, we use either $\bar{w}$ or $\hat{w}$. If F is symmetric, then

$$\hat{\tau} = (\bar{w}_p - \bar{w}_q)/2F^{-1}(p).$$

An analogous estimate in terms of $\hat{w}$ holds. In the case of the logistic F,

$$\hat{\tau} = (\bar{w}_p - \bar{w}_q)/2\ln(p/q).$$

8.6. Expected Values of the Sample Size

Wetherill et al. [1966] give some discussion and methods based on the Markov chain approach for computing the average sample number. For further details, see Wetherill et al. [1966, section 12].

8.7. Up and Down Methods with Multiple Sampling

So far we have discussed up and down methods with an observation at each stage. Tsutakawa [1967] proposed up and down methods with data obtained sequentially in blocks. If $L = \{\ldots, d_{-1}d_0, d_1, \ldots\}$ is a set of equally spaced levels with interval d, the experiment starts with K observations at some level Y_1 in L and continues for an additional $n - 1$ trials at levels $Y_2, \ldots, Y_n$ determined by

$$Y_{i+1} = \begin{cases} Y_i + d & \text{if } 0 \leqslant r_i \leqslant k^0 \\ Y_i & \text{if } k^0 < r_i < K - k^0 \\ Y_i - d & \text{if } K - k_0 \leqslant r_i \leqslant K \end{cases}$$

where r_i denotes the number of responses at the i-th trial and k^0 is a specified integer such that $0 \leqslant k^0 < K - k^0$. (Notice that, for given $Y_i = d_j$, r_i is binominally distributed with probability of response $F(d_j)$ for each observation.) The procedure for generating the sequence $\{Y_i, r_i, i \geqslant 1\}$ will be called the random walk design and is denoted by $W(K, k^0)$. The design $W(1, 0)$ is the up and down method of Dixon and Mood [1948].

Using the results of the Markov chain theory, the asymptotic distribution of the sample average, $\bar{y}$ of the dose levels, is studied and an estimate of its variance is proposed by Tsutakawa [1967]. Numerical studies of the bias and mean square error of $\bar{y}$ indicate that there is often a loss of efficiency in using K up to 5 instead of $K = 1$. Tsutakawa [1967] also surmises from the asymptotic properties of the Spearman-Karber estimator, that its large sample efficiency can decrease as the number of distinct levels increases, whereas in small samples, it is quite robust.

Hsi [1969] proposed an up and down method using multiple samples (MUD). The procedure is as follows:

(1) A series of dose levels is chosen, usually in units of log-doses.

(2) A sequence of trials, using k experimental units at a time is carried out. At each trial the dosage level is selected depending upon the proportion of units responding, as follows:

(a) Go to the next higher dose level if s or less responses among k units are found in the current trial;

(b) go to the lower dose level if r or more out of k respond at the current level ($r > s$);

(c) use the same dose level if the number of responses lies strictly between s and r.

(3) The experiment is terminated after n trials and the estimator is computed (usually n is prespecified). Apparently the MUD procedure is slightly inefficient when compared with the sequential up and down method (SUD method), especially if the initial dose level is far away from the percentage dose to be estimated.

Hsi [1969] computes the bias and precision of the estimators for LD_{50} based on the dose-averaging formula for the probit model. A method of estimating the extreme percentage doses is also indicated. The MUD method is more efficient than the nonsequential procedures and is comparable to other SUDs.

8.8. Estimation of Extreme Quantiles

The estimation of low dose levels is of much interest, especially in carcinogenic models. Classical sampling procedures involve placing the doses into equally sized k values, equally spaced groups and then using maximum likelihood methods for estimating the unknown parameters. Sequential methods call for selecting a dose and observing the response before the next dose is administered. Let Y_1 be the response at dose level x_1 given to a subject and we administer dose x_2 to the second subject. The same point estimation procedures that are used in a fixed dose level case may also be used in the sequential case; however, the dose levels are randomly selected. Typically x_{n+1} depends on $x_1, \ldots, x_n$ and the respective responses $Y_1, \ldots, Y_n$. These were discussed in earlier sections of this chapter. McLeish and Tosh [1983] propose a sequential method for logit analysis for the following reasons:

(1) The sequential method is reasonably robust against changes in response function form;

(2) it allows other forms of estimation which are easier to handle than the method of maximum likelihood;

(3) it selects the dose levels in such a way that more information pertaining to the extreme quantiles of interest from the data is gained while controlling the number of deaths of experimental units.

Note that property (3) is important in which the major control of a study is associated with the number of deaths among the test units as a result of an experiment, for example, in studies that involve the use of higher mammals such as chimpanzees.

8.8.1. Sequential Procedure of McLeish and Tosh [1983]

Let

$$P[Y(x) = 1] = F(x) = [1 + e^{-\beta(x-\theta)}]^{-1} \tag{1}$$

where β is the scale parameter and θ is the location parameter. We will be interested in estimating the root r of the equation

$$F(x) = p, \tag{2}$$

when p is close to zero. For instance, LD_5 is the root of $F(x) = 0.05$.

Observations will be made at dose levels $x_1, x_2, \ldots, x_n$ until the desired response is observed. Let $Y_1, Y_2, \ldots, Y_n$ be the responses corresponding to $x_1, x_2, \ldots, x_n$ and let $\Delta > 0$ be given. Then for estimating a root r of (2) when p is near 0, the procedure is to sample sequentially at

$$x_1, x_2 = x_1 + \Delta, \ldots, x_N = x_1 + (N-1)\Delta, \quad \text{where} \tag{3}$$

$$N = \inf\{j: Y_j = 1\}. \tag{4}$$

Estimates will be based on x_1, Δ and N or x_1, Δ and the range of doses $(N-1)\Delta$. After adding a standard continuity correction factor $\Delta/2$, define $D = (N - \frac{1}{2})\Delta$.

8.8.2. Estimation of Parameters

We now investigate the estimation of parameters based on the exponential approximation (By the theorem in Appendix, for $\Delta \to 0$ and $x_1 \to -\infty$, $[\exp(\beta D) - 1]$ is approximately negative exponential with mean $1/m$ where $m = e^{\beta(x_1-\theta)}/[e^{\beta\Delta} - 1]$.) Hereafter the 'phrases'-like maximum likelihood (unbiased) will be based on this approximating density. For k realizations of D, $d_1, \ldots, d_k$, starting points $x_{11}, \ldots, x_{1k}$ increment sizes $\Delta_1, \ldots, \Delta_k$ and corresponding values of $m_1, \ldots, m_k$, the mle values of θ and β are obtained from the log likelihood

$$\ln L = k \ln \beta + \sum_{j=1}^{k} \ln m_j + \beta \sum_{j=1}^{k} d_j - \sum_{j=1}^{k} m_j (e^{\beta d_j} - 1).$$

However, one cannot obtain explicit solutions. When β is known and $x_{ij} \equiv x_1$, $\Delta_i \equiv \Delta$, the authors claim that the mle of θ differs from the uniformly minimum variance unbiased estimator of θ by the constant function $[\psi(k) - \ln k]/\beta$ where ψ denotes the digamma function $[\psi(x) = d \ln \Gamma(x)/dx]$. Further, moment estimates of β and θ based on the extreme value distribution approximation are

$$\tilde{\beta} = \pi[6 \operatorname{svar}(x_n)]^{-1/2}, \quad \tilde{\theta} = \operatorname{av}\{x_N + [\gamma - \ln(e^{\tilde{\beta}\Delta} - 1)]/\tilde{\beta}\},$$

where svar denotes sample variance, $\operatorname{av}(\cdot)$ denotes mean and γ is Euler's constant. The above estimates of β, θ and hence of r have considerable bias, but small variance.

Remark. The convergence to an exponential distribution is valid when $F(x) \sim c \exp(-\alpha x)$ as $x \to -\infty$, which is satisfied by a broad class of distributions, but fails when F is normal. However, the authors find the first response estimates, namely $\tilde{\beta}$ and $\tilde{\theta}$ to be good when applied to the probit model.
The estimation scheme is strictly sequential, that is, Y_n must be observed before dose x_{n+1} is administered. The authors suggest the following remedy. For example, we plan to continue sampling until we observe 10 deaths. Then administer dose x_1 to 10 test units and observe the number z_1 of deaths. Next, we administer dose $x_2 = x_1 + \Delta$ to $10 - z_1$ units and observe the number z_2 deaths. Then administer dose $x_3 = x_2 + \Delta$ to $10 - z_1 - z_2$ units and observe z_3; continue in this manner until a total of 10 deaths is observed. The 10 values of N are the indices of the doses x_j at which the deaths were recorded. The above procedure is simply a more efficient way of obtaining 10 replicate values of the variable D.

8.8.3. Choice of x_1 and Δ

Based on the exponential approximation given in the appendix and the cost of administering a dose being 1 and the additional cost of death being C_1, the authors propose a method of optimally choosing x_1 and Δ by minimizing a certain criterion function, namely the product of the asymptotic variance of $\hat{r}$ (based on the information matrix) and $[E(N) + C_1]$. For further details see McLeish and Tosh [1983, section 4].

Appendix: Limiting Distribution of $D = (N - 1/2)\Delta$

The following result is used by McLeish and Tosh [1983].

Theorem

As $x_1 \to -\infty$ and $\Delta \to 0$, $m(e^{\beta D} - 1)$ converges in distribution to a standard exponential variable where

$$m = e^{\beta(x_1 - \theta)}/(e^{\beta\Delta} - 1).$$

Proof. For arbitrary $\xi > 0$, define d and the integer n by

$$d = \beta^{-1} \ln(1 + \xi/m) \quad (0 \leq n\Delta - d < \Delta).$$

Now, if $x_1 \to -\infty$ and $\Delta \to 0$, $n \to \infty$ provided d is bounded away from zero. However, if a subsequence of d values approaches zero, for this subsequence

$$d \sim \beta^{-1}\xi/m - \xi\Delta/e^{\beta(x_1 - \theta)}$$

and consequently $d/\Delta \to \infty$. So, in any case $n \to \infty$ and let us approximately set $D = N\Delta$. Next, since $(n-1)\Delta < d < n\Delta$,

$$P(N > n - 1) > P(N > d) > P(N > n), \quad \text{hence}$$

$$-\ln P(N > n - 1) < -\ln P(D > d) < -\ln P(N > n).$$

Also we have the inequality $a - a^2 \leq \ln(1 + a) \leq a$ for $a > 0$. Hence

$$-\ln P(D > d) \leq \sum_{i=1}^{n} e^{\beta(x_i - \theta)} = e^{\beta(x_1 - \theta)} \sum_{i=1}^{n} e^{(i-1)\Delta\beta} \leq m(e^{n\beta\Delta} - 1).$$

Similarly, one can obtain

$$-\ln P(D > d) \geq m(e^{\beta n\Delta} - 1) - m^2(e^{2\beta n\Delta} - 1)(e^{\beta\Delta} - 1)/(e^{\beta\Delta} + 1).$$

Now, the second term on the right-hand side goes to zero since it is nonnegative and bounded above by

$$(e^{\beta\Delta} + 1)e^{2\beta(x_1 - \theta)} + 2e^{\beta(x_1 - \theta + 2\Delta)} + \xi^2 e^{2\beta\Delta}(e^{\beta\Delta} - 1),$$

which approaches 0 as $x_1 \to -\infty$ and $\Delta \to 0$. Also note that under the same conditions

$$(e^{\beta n\Delta} - 1)/(e^{\beta d} - 1) \to 1.$$

Hence, applying the above approximations, we have

$$-\ln P(D > d) = -\ln P[m(e^{\beta D} - 1) > (e^{\beta d} - 1)m]$$

$$= -\ln P[m(e^{\beta D} - 1) > \xi] \to \xi,$$

as $x_1 \to -\infty$ and $\Delta \to 0$, which completes the proof of the theorem.
Next, if $G(x)$ denotes the extreme value survivor function

$$G(x) = \exp[-e^{\beta(x-\psi)}],$$

where $\psi = \theta + \beta^{-1}\ln(e^{\beta\Delta} - 1)$, we have, from the exponential approximation for $x > x_1$,

$$P(x_N > x) = P(D > x - x_1) \doteq G(x)/G(x_1),$$

which is the survivor function of a truncated extreme value distribution. The mass truncated is $1 - G(x_1) = 1 - e^{-m}$. Hence, if $x_1 \to -\infty$ and $\Delta \to 0$ in such a way that $m \to 0$, then $x_N - \theta - \beta^{-1}\ln(e^{\beta\Delta} - 1)$ converges in distribution to an extreme value distribution with location zero and scale β. Furthermore, from the theorem, we compute the density of D to be

$$f(d) = m\beta e^{\beta d} \exp[-m(e^{\beta d} - 1)].$$

9 Estimation of 'Safe Doses'

The 'Delaney Clause' published in 1958, as part of the Food Additive Amendments to the Food, Drug, and Cosmetic Act, says that if a substance is found to induce cancer in man or animal, after 'appropriate' experimental testing, then this substance may not be used as an additive in food. The clause requires unconditional banning of the use of a substance found to induce cancer at extremely high dose levels while ignoring possible beneficial effects available at much lower 'use' levels where the carcinogenic risk might be very minimal. This raises questions regarding the definition and determination of 'safe' residual doses for potential cancer-inducing substances. A dose of a substance is currently said to be 'safe' if the induced cancer rate does not differ appreciably from the zero dose cancer incidence rate.

9.1. *Models for Carcinogenic Rates*

Models for carcinogenic hazard rates have been primarily proposed by Mantel and Bryan [1961], Armitage and Doll [1961], Peto and Lee [1973] and Hartley and Sielkin [1977]. Here we shall present the model considered by Hartley and Sielkin [1977]. Although in some experiments the time to tumor is available, the model of Hartley and Sielkin [1977] covers situations in which for some or all animals the time to tumor is not available, but it is only known whether or not the animal developed cancer before the time of sacrifice or before death. Let

$$F(t,x;\alpha,\beta) = P \text{ (the time to tumor is less than or equal to t, the dose is x)} \quad (9.1)$$

where α and β denote vector parameters. Also let

$$f(t,x;\alpha,\beta) = dF(t,x,\alpha,\beta)/dt \quad (9.2)$$

$$\bar{F}(t,x;\alpha,\beta) = 1 - F(t,x;\alpha,\beta). \quad (9.3)$$

Let the age-specific tumor incidence rate or hazard rate be denoted by

$$H(t,x;\alpha,\beta) = f(t,x;\alpha,\beta)/\bar{F}(t,x;\alpha,\beta) = -\frac{d}{dt}\ln \bar{F}(t,x;\alpha,\beta). \quad (9.4)$$

If only the tumor incidence counts are available at the termination T of the experiment, we use the symbol $F(x) = F(T,x;\alpha,\beta)$ to denote dependence of the tumor incidence rate on the dose level.

The product model for the hazard rate H stipulates that H can be factorized as follows

$$H(t,x;\alpha,\beta) = g(x;\alpha)l(t;\beta). \tag{9.5}$$

This model is acceptable for experiments in which an agent is applied at a constant rate continuously over time. Cox [1972] maintains that $g(x;\alpha)$ should be completely specified and $l(t;\beta)$ need not be specified, that is, choose a non-parametric form for $l(t;\beta)$. If α is a parameter of prime importance, then $l(t;\beta)$ is a 'nuisance function'. However, in carcinogenic safety testing the definition of safe dose requires the estimation of both $l(t;\beta)$ and $g(x;\alpha)$ although the inferences may be more sensitive to the precise form of the function $g(x;\alpha)$ than to that of $l(t;\beta)$.

Hartley and Sielkin [1977] assume that

$$L(t;\beta) = \int_0^t l(\tau;\beta)d\tau = \sum_{r=1}^b \beta_r t^r \quad (\beta_r \geq 0) \tag{9.6}$$

where $\beta_0 = 0$, since at $t = 0$ the hazard rate must be zero and where without loss of generality we can standardize the coefficients so that

$$L(1;\beta) = \sum_{r=1}^b \beta_r = 1. \tag{9.7}$$

The polynomial form (9.6) can be regarded as a weighted average of Weibull hazard rates with positive weight coefficients β_r. Since positive polynomials constitute a system for representing any continuous function, the model considered by Cox [1972] can be construed as a limiting case of (9.6) with $b \to \infty$.

Further, we assume that

$$g(x;\alpha) = \sum_{s=0}^a \alpha_s x^s, \quad \text{where} \tag{9.8}$$

$$\alpha_s \geq 0 \quad \text{(hence } g(x;\alpha) \geq 0 \text{ and } d^2 g(x;\alpha)/dx^2 \geq 0). \tag{9.9}$$

Note that Armitage and Doll [1961] take $g(x;\alpha)$ to be

$$g(x;\alpha) = c\prod_{s=1}^a (1 + \alpha_s x). \tag{9.10}$$

9.2. *Maximum Likelihood Estimation of the Parameters*

Let D denote the number of dose levels and n_d denote the number of experimental units at dose level x_d. In order to write down the likelihood function, the following assumptions will be made. Each experimental unit is associated with (1) the dose x_d, (2) the prespecified time T at which it will be destroyed, (3) the time to cancer L if observable, and (4) the time of death τ provided $\tau \leq T$.

Computation of Likelihood under Various Experimental Data

(1) We observe time to cancer t without necropsy (e.g. palpability). Then t is a random variable from uncensored range $t \leq T$. The associated component of the likelihood is $f(t,x;\alpha,\beta)$.

(2) There is negative necropsy at termination T. Then time to cancer occurs some time after T. Hence the associated component of the likelihood is $\bar{F}(T,x;\alpha,\beta)$.

(3) There is negative necropsy at death τ and the death is due to causes independent of cancer; t is censored at random value τ from the life distribution. Here the factor of likelihood is $\bar{F}(\tau,x;\alpha,\beta)$.

(4) There is positive necropsy at termination T or at death τ; t can be estimated from the necropsy information. The estimated value of t, denoted by t′, is equivalent to random observation from the time to cancer distribution. Then the likelihood factor is $f(t',x;\alpha,\beta)$.

(5) Positive necropsy at termination T or death τ; however, $t \leq T$ or $t \leq \tau$ is all that is known about t. Since t cannot be estimated at necropsy, it is random in the interval (0,T) or (0,τ). Consequently, the likelihood component is $\hat{F}(T,x;\alpha,\beta)$ or $F(\tau,x;\alpha,\beta)$.

In experiments in which only counts of experimental units with or without cancer are made, we would be confined to data types 2, 3 and 5. In the computation of the likelihood, T is used equivalently to τ and t′ equivalently to t. The n_d units are divided into three categories with

n_{d1} = the number of experimental units for which the time to tumor is either recorded or estimated on the basis of a positive necropsy (data types 1 and 4)

n_{d2} = the number of units with a positive necropsy, but no known time to tumor or estimated time to tumor (data type 5), and

n_{d3} = the number of units with a negative necropsy (data types 2 and 3).

Then the likelihood of α and β is

$$L = \prod_{d=1}^{D} \left[\prod_{i=1}^{n_{d1}} f(t_i) \prod_{j=1}^{n_{d2}} F(\tau_j) \prod_{k=1}^{n_{d3}} \bar{F}(\tau_k) \right], \quad \text{where} \tag{9.11}$$

$F(t,x;\alpha,\beta) = 1 - \exp[-g(x;\alpha)L(t;\beta)]$

$\bar{F}(t,x;\alpha,\beta) = 1 - F(t,x;\alpha,\beta)$

$f(t,x;\alpha,\beta) = g(x;\alpha)l(t;\beta)\exp[-g(x;\alpha)L(t;\beta)]$

$$L(t;\beta) = \sum_{r=1}^{b} \beta_r t^r. \tag{9.12}$$

Maximum likelihood estimators (mle) $\hat{\alpha}$, $\hat{\beta}$ can now be computed by the iterative convex programming algorithms, which will be described below.

9.3. Convex Programming Algorithms

We would like to find $\hat{\alpha}$ and $\hat{\beta}$ that would minimize $\Delta = -\ln L$. For given β, the problem of minimizing Δ, with respect to α subject to $\alpha_s \geq 0$ or constraints in (9.10), can be handled by a suitable convex programing algorithm leading to a unique global minimum of Δ. Similarly, we can handle the problem of minimizing Δ for given α as a function of β subject to $\beta_r \geq 0$ or subject to the linear inequality constraints $l(t;\beta) \geq 0$.

Let $\alpha(0)$, $\beta(0)$ be the initial consistent estimates.

Step 1. Minimize $\Delta[\alpha,\beta(0)]$ with respect to α subject to linear constraints and let $\alpha(1)$ be the minimizing solution.

Step 2. Minimize $\Delta[\alpha(1),\beta]$ with respect to β subject to the linear constraints and let $\beta(1)$ be the solution.

For moderate values of a and b steps 1 and 2 consist of simple convex programming problems, especially if the simple constraints $\alpha_s \geq 0$, $\beta_r \geq 0$ and (9.7) are used. In step 1, the constraints $\alpha_s \geq 0$ are automatically satisfied, and in step 2 only one linear constraint (9.7) need to be met. Thus, Δ will be nonincreasing at every cycle step and hence will converge to a lower limit Δ^*. Since we introduce upper bounds for the α_s, β_r, the generated sequence $\alpha(I)$, $\beta(I)$ (I denoting the cycle) will have at least one point of accumulation. Now, the Kuhn-Tucker equations are

$$\left. \begin{array}{l} -\partial\Delta/\partial\alpha_s + \lambda_s = 0 \\ \alpha_s \geq 0, \lambda_s\alpha_s = 0; \lambda_s \geq 0 \end{array} \right\} \quad s = 0,\ldots,a \tag{9.13}$$

$$\left. \begin{array}{l} -\partial\Delta/\partial\beta_r + \mu_r + \mu = 0 \\ \beta_r \geq 0; \mu_r\beta_r = 0; \mu_r \geq 0 \end{array} \right\} \quad r = 1,\ldots,b.$$

$$\mu\left(1 - \sum_{r=1}^{b} \beta_r\right) = 0 \tag{9.14}$$

because the step 1 and step 2 problems are both satisfied for the step 1 and step 2 problems at every point of accumulation and the objective function has the same value Δ^* at every point of accumulation. Hence, if we assume that equations (9.13) and (9.14) cannot have two different solution points α, β with the same value of the objective function, then there can be only one point of accumulation; therefore

$$\lim_{I\to\infty}\alpha(I) = \hat{\alpha}, \quad \lim_{I\to\infty}\beta(I) = \hat{\beta}.$$

That is, the two-step iteration process converges to a solution of equations (9.13) and (9.14).

Procedure for Obtaining the Consistent Estimators α(0), β(0)

The cancer data are split into D separate dose groups, and using the convex programming algorithm of step 2 separately for the data in each group omitting the restriction (9.7) and using the values $\alpha_0 = 1$, $\alpha_s = 0$, $s = 1, \ldots, a$, we obtain estimators $\tilde{\beta}_{r,d}$ for each group. These estimators are global mle values of the true parameter functions $\beta_r \left[\sum_{s=0}^{a} \alpha_s x_d^s \right]$ at dose x_d, and they are consistent. Next, we compute for each d the quantities $B_d = \sum_{r=1}^{b} \tilde{\beta}_{r,d}$ which are consistent estimators of the parameters

$$\left(\sum_{r=1}^{b} \beta_r \right) \left(\sum_{s=0}^{a} \alpha_s x_d^s \right) = \sum_{s=0}^{a} \alpha_s x_d^s.$$

Next, we compute the unweighted least-square estimators $\alpha_s(0)$ by fitting the form (9.8) to B_d. The $\alpha_s(0)$ are then consistent estimators of the α_s since the Vandermonde matrix (x_d^s) does not depend on n. Finally, the consistent estimators $\beta_r(0)$ are computed from

$$\beta_r(0) = \frac{1}{D} \sum_{d=1}^{D} \hat{\beta}_{rd} \bigg/ \left[\sum_{s=0}^{a} \alpha_s(0) x_d^s \right]. \tag{9.15}$$

Thus, the above method involves D applications of the simple reduced step 2 convex programming algorithm and a least-squares computation.

There are basically three nonstandard features involved: namely,
(1) the time to tumor data is in general, partially or totally, censored;
(2) the varying dose levels make the observation not identically distributed;
(3) the parameter space is restricted by linear inequalities.

Much progress has been made in recent years to obtain mle values subject to constraints.

The following property can easily be verified: the matrix of second derivatives of Δ with respect to α is always positive semidefinite and strictly positive definite if (1) there is at least one time to tumor observation t_{d_i}, for each dose and/or at least one τ_{d_i} record is available for each dose, and (2) the number of doses is larger than a. Similarly, the matrix of second derivatives of Δ with respect to β is always positive semidefinite and is strictly positive definite for at least one dose, d. There are at least b different values of τ_{d_i} or t_{d_i}. The above conditions assure the uniqueness of the mle $\hat{\alpha}$, $\hat{\beta}$.

9.4. *Point Estimation and Confidence Intervals for 'Safe Doses'*

A 'safe dose' ξ specifies a tolerable increment in the cancer rate over the spontaneous cancer rate. For example, let $\Pi = 10^{-8}$ and then define ξ by the equation

$$\Pi = F(T^*, \xi; \alpha, \beta) - F(T^*, 0; \alpha, \beta) \tag{9.16}$$

where α, β are the true parameter vectors and T^* is a conveniently chosen 'exposure time' (typically the duration of the experiment). Equation (9.16) implies that 'tolerance', $\text{Tol}(T^*)$, which is a permissible percentage reduction in the spontaneous tumor-free proportion, is given by

$$\text{Tol}(T^*) = \bar{F}(T^*, \xi; \alpha, \beta)/\bar{F}(T^*, 0; \alpha, \beta) = 1 - \Pi[\bar{F}(T^*, 0; \alpha, \beta)]^{-1}. \tag{9.17}$$

However, if $\text{Tol}(T^*)$ rather than Π is specified, then ξ is the solution of the equation

$$[-\ln \text{Tol}(T^*)]\left(\sum_{r=1}^{b} \beta_r T^{*r}\right)^{-1} = \sum_{s=1}^{a} \alpha_s \xi^s. \tag{9.18}$$

The mle of ξ is obtained by replacing β_r by $\hat{\beta}_r$ and α_s by $\hat{\alpha}_s$, (the respective mle). Obtaining a lower confidence limit for ξ is more difficult since the asymptotic covariance matrix of $\hat{\alpha}$, $\hat{\beta}$ is not known. Hartley and Sielkin [1977] propose a more direct approach.

Split the data randomly into G groups obtaining separate mle $\hat{\alpha}_s(g)$, $\hat{\beta}_r(g)$ for each group and determining the safe dose estimate $x(g)$ from the equations

$$[-\ln \text{Tol}(T^*)]\left[\sum_{s=1}^{b} \hat{\beta}_r(g) T^{*r}\right]^{-1} = \sum_{s=1}^{a} \hat{\alpha}_s(g) \hat{x}(g)^s, \quad g = 1, \ldots, G. \tag{9.19}$$

Now, assuming an approximate normal distribution for $z(g) = \ln \hat{x}(g)$, lower confidence limits z_1 can be computed from the equation

$$z_1 = \ln x_1 = \bar{z} - t_{0.95, G-1}\left\{\sum_{g=1}^{G} \frac{[z(g) - \bar{z}]^2}{G(G-1)}\right\}^{1/2}. \tag{9.20}$$

The above procedure may be biased even for moderate sample sizes. Hence, the authors suggest replacing $\bar{z}$ by $\ln \hat{\xi}$ where

$$\hat{\xi} = [-\ln \text{Tol}(T^*)]/\bar{u}; \quad \bar{u} = G^{-1} \sum_{g=1}^{G}\left[\sum_{s=1}^{b} \hat{\beta}_r(g) T^{*r}\right] \hat{\alpha}_1(g) \tag{9.21}$$

If $\hat{\alpha}_s(g) = 0$ for $g = 1, \ldots, G$ and $s = 1, \ldots, s' - 1$, then

$$\hat{\xi} = [-\ln \text{Tol}(T^*)]^{-1/s'}/\bar{u}_{s'};$$

$$\bar{u}_{s'} = G^{-1} \sum_{g=1}^{G}\left\{\sum_{s=1}^{b} \hat{\beta}_r(g) T^{*r}\right\} \hat{\alpha}_{s'}(g).$$

Example

Hartley and Sielkin [1977] analyzed the data of Dr. D. W. Gayor of the Biometry Division of the National Center for Toxicological Research, which consists of two sets of tumor-rate data. In each set there were $D = 3$ non-zero dosage levels.

Doses of 2-acetylamine fluorene (2-AAF) were fed to a specified strain of mice. The spontaneous rate of bladder tumor was given as zero. Times to tumor were taken to be the number of days from weaning (21 days) of tumor incidence. In each of the two data sets the observations at each dosage level were randomly subdivided into $G = 6$ groups of approximately equal size, here $T^* = 550$ which was approximately the termination time of the experiment. The $\text{Tol}(T^*) = 0.9999$. In the parametric Weibull model (PWM) $a = 3$, and in both PWM and the nonparametric Weibull model (NPWM) $\hat{k} = 0.2, 0.4, \ldots, 8.0$. In the general product model (GPM) $a = 3$ and $b = 8$. Also, since the probability of cancer at dose $x = 0$ was given as zero, we set $\alpha_0 = 0$ in PWM and GPM while $\gamma_0 = 0$ in NPWM. (See (9.8), (9.12) and page 124 for these models.)

Table 9.1 contains the observed tumor frequency, the estimated safe dose $\hat{\xi}$ and the lower 95% confidence limit x_f on the true safe dose ξ for each data set.

Table 9.1. Life data and its analysis

Data set	Number of mice	Dose level	Developing cancer, %
1	137	100	3
	130	250	45
	121	500	64
2	211	100	27
	39	250	100
	125	500	92

Data set	Estimated safe dose			Lower confidence limit x_1		
	PWM	NWPM	GPM	PWM	NPWM	GPM
1	0.71	0.21	1.1	0.17	0.01	0.31
2	0.08	0.03	0.07	0.01	0.02	0.01

9.5. *The Mantel-Bryan Model*

Let $Y(x)$ denote the probit transform given by

$$F(x) - L = (1 - L)\Phi(Y(x)) \tag{9.22}$$

where L denotes the spontaneous cancer rate. For the conservative procedure, Mantel and Bryan [1961] assume that whatever form the true response curve may take over the region of extrapolation, the average slope is not less than the assumed one.

However, by setting $z = \ln x$, the above probit model should satisfy

$$(z_2 - z_1)^{-1} \int_{z_1}^{z_2} \frac{dY}{dz} dz \geq c \quad \text{or} \tag{9.23}$$

$$|Y(z_2) - Y(z_1)| \geq (z_2 - z_1)c \tag{9.24}$$

where z_2 is the log dose from which downward extrapolation is made, and z_1 is the log dose to which extrapolation is made. Since we are concerned with extremely low doses, we must note that the range of extrapolation is $-\infty < z_1 < z_2$. Hartley and Sielkin [1977] show that if $F(x)$ is analytic at $x = 0$, the inequalities (9.23) and (9.24) may be violated for small doses x_i. For instance, if $F(x) - L \doteq a_v x^v$ ($a_v > 0, v \geq 1$) for small x, then from (9.22) and using the approximation

$$\Phi(x) = \phi(x)/|x| \quad \text{for small x}$$

we have

$$\ln x_1 = z_1 \doteq \ln\{[F(z_1) - L]/a_v\}^{1/v}$$
$$\doteq -\ln a_v/v + v^{-1}\ln\{(1 - L)(2\pi)^{-1/2}\exp[-Y^2(z_1)/2]|Y(z_1)|^{-1}\}$$
$$\doteq [\text{const} - Y^2(z_1)/2 - \ln|Y(z_1)|]/v.$$

Substituting this for z_1 in (9.24) we see that (9.24) will not be satisfied as $Y(z_1) \to -\infty$.

The authors carry out Monte Carlo and real data studies on the model considered here, the PWM and the NPWM given, respectively, by

$$F(t,x;\alpha,\beta) = 1 - \exp\left[-\left(\sum_{s=0}^{a} \alpha_s x^s\right) t^k\right], \quad \alpha_s \geq 0, k > 0, \quad \text{and}$$

$$F(t,x_d;\alpha,\beta) = 1 - \exp(-\gamma_d t^k), \quad d = 1, \ldots, D$$

$$0 \leq \gamma_1 \leq \gamma_2 \leq, \ldots, \leq \gamma_D$$

$$\frac{\gamma_{d+2} - \gamma_{d+1}}{x_{d+2} - \alpha_{d+1}} \geq \frac{\gamma_{d+1} - \gamma_d}{x_{d+1} - x_d}, \quad d = 1, \ldots, D-2$$

where γ_d is the nonparametric representation of $g(x_d;\alpha)$. The simulation study indicates that the safe dose estimates are robust to changes in the model for $L(t;\beta)$. For estimation of ξ, the parametric models perform better than the nonparametric model.

9.6. *Dose-Response Relationships Based on Dichotomous Data*

Due to the number of chemicals which must be tested, the number of experimental units that are used in any one experiment is small. Hence, the dose levels should be set high enough to produce tumors in an appreciable fraction (for example,

10^{-1} or more) of the experimental units. The statistical problem is to use these high-dose data to estimate the dose level at which the cancer risk in the units would exceed the background cancer risk (that is, the risk at zero dose) by no more than some specified low amount, say 10^{-6}. This problem is commonly known as the *low-dose extrapolation* problem. Crump et al. [1977] address themselves to this problem when the data is dichotomous with a model different from that of Hartley and Sielkin [1977].

9.6.1. Description of the Model

Any low-dose extrapolation procedure depends strongly on the assumed dose-response relationship. Many dose-response models provide reasonably good fits to the data in the experimental dose range, but yield risk estimates that differ by several orders of magnitude in the low-dose range. Some models such as the one-hit or one-stage model of Armitage and Doll [1961] assume that the risk is approximately linear in dose in the low-dose range. Also, the probit model of Mantel and Bryan [1961] and Mantel et al. [1957] has a dose-response function which has derivatives of all orders approaching zero as the dose level tends to zero. Thus, the dose-response function is extremely flat in the very low-dose region. As one alternative, Crump et al. [1977] consider the dose-response function of the form

$$F(d) = 1 - \exp(-Q(d)) \tag{9.25}$$

where $Q(s)$ is a polynomial with nonnegative coefficients and of known degree, that is

$$Q(D) = \sum_{i=0}^{\infty} q_i d^i \quad (q_i \geq 0 \text{ for all } i) \tag{9.26}$$

and $q_i = 0$ for all but a finite number of i. If $q_1 > 0$, $F(d)$ is linear in d for small d. On the other hand $F(d)$ can be made arbitrarily flat by taking a sufficient number of the low-order coefficients equal to zero, because at very low doses $F(d) - F(0) \doteq \exp(-q_0) q_l d^l$ where l is the smallest positive integer for which $q_l > 0$. Notice that $F(d)$ denotes the probability that a unit will develop a particular type of tumor during the duration of the experiment while under continuous exposure to a dose d of the drug.

Point estimates of the coefficients q_i (i = 0, 1, ...) and hence the response probabilities $F(d)$ for any dose level d can be obtained by maximizing the likelihood function of the data. When the degree of the polynomial Q is unknown, we need to estimate an infinite number of the q_i. Conditions for existence and uniqueness of solutions to this infinite parameter maximization problem are given by Guess and Crump [1978]. The likelihood L does not achieve a maximum within the class of functions determined by (9.26). However, Guess and Crump [1978] show that the likelihood will achieve a maximum if the class of functions over which the maximum is taken is enlarged so as to include polynomials of infinite degree. That is, if we redefine Q as

$$Q(d) = \sum_{i=0}^{\infty} q_i d^i + q_\infty d^\infty, \quad q_i \geq 0 \quad (i = 0, 1, \ldots) \qquad (9.27)$$

where we define $d^\infty = 1$ if $d \geq d_n$ and zero otherwise (d_n equals the highest test dose appearing in the likelihood function) and where all but finitely many of the q_i are zero. The inclusion of the step function assures the existence of a maximum likelihood solution. The procedure for computing the function Q of the form (9.27) that maximizes the likelihood is called a global maximization. The asymptotic distribution of the maximum likelihood coefficient vector $\hat{q}$ is derived by Guess and Crump [1978]. Because of the nonnegative constraints $q_i \geq 0$, the asymptotic distribution of $\hat{q}$ is not multinormal [Crump et al., 1977, p. 450]. Crump et al. [1977] compute confidence intervals for P(d) assuming that there exists a positive integer J such that $q_i > 0$ for $0 \leq i \leq J$ and $q_i = 0$ for $i > J$ (because then, the usual asymptotic multinormality of $\hat{q}$ and hence of $\hat{F}(d)$ is valid). The simulation studies indicate that the asymptotic confidence intervals are nearly correct.

9.6.2. Testing of Hypotheses

The authors construct likelihood ratio test procedures for H_{01}: $q_0 > 0, q_1 = \cdots = q_J = 0$ versus H_{11}: $q_0 > 0, q_j \geq 0$ for $j = 1, \ldots, J, q_i > 0$ for some $J = 1, \ldots, J$, for H_{02}: $q_1 = 0, q_j > 0$ for all $j = 0, 2, \ldots, J$ versus H_{12}: $q_i > 0$ for all $j = 0, 1, \ldots, J$. Notice that when $J = 1$, the two test procedures coincide. Since nonparametric tests are available [Barlow et al., 1972, pp. 192–194] for H_{01} versus H_{11}, Crump et al. [1977] focus their attention on H_{02} versus H_{12}. They conduct a number of simulation experiments. For the results of the simulation, refer to table 2 of Crump et al. [1977] who also carry out some Monte Carlo goodness-of-fit studies.

The asymptotic theory of the estimate $\hat{q}$, $\hat{F}(d)$ and $\hat{d}_a$ for some small a ($0 < a$) is given in appendix 1 of Crump et al. [1977], where d_a is such that $F(d_a) - F(0) = \Pi_a$ and Π_a is a known fixed value.

Remark. The approach of Hartley and Sielkin [1977] utilizes time to tumor data whereas the approach of Crump et al. [1977] does not. However, when the data is dichotomous, the mle for the dose that yields a given risk should be the same by both procedures. However, the methods are different for obtaining the asymptotic confidence intervals.

Crump et al. [1977] fit the polynomial Q(D) to certain data and part of their numerical study is presented in table 9.2.

Table 9.2. Experimental data and estimates of Q(D) obtained by global maximization of the likelihood function

Dose level ppm	Number of responders/ number of animals tested	Estimated coefficients from a global fit to data
Dieldrin [Walker et al., 1972]		
0.00	17/156	$\hat{q}_0 = .113$
1.25	11/60	$\hat{q}_1 = .051$
2.50	25/58	$\hat{q}_2 = .040$
5.00[1]	44/60	$\hat{q}_i = 0, \; i > 2$
DDT [Tomatis et al., 1972]		
0	4/11	$\hat{q}_0 = .045$
2	4/105	$\hat{q}_1 = .002$
10	11/124	$\hat{q}_\infty = .541$
50	13/104	$\hat{q}_i = 0$, otherwise
100	60/90	

[1] Data at higher dose levels are omitted.

Crump et al. [1977, p. 443] have graphed upper confidence intervals for added risk and lower confidence intervals on dose for the data sets in table 9.2.

9.7. *Optimal Designs in Carcinogen Experiments*

Optimal designs associated with estimating response probabilities in low-dose carcinogen are of much interest. Hoel and Jennrich [1979] develop optimal designs for a specific class of models and study their robustness with respect to the prior information regarding the model. They also provide an algorithm for constructing the optimal design.

Regression with Nonconstant Variance

Let $f_0(x), \ldots, f_k(x)$ be a Chebyshev system of linearly independent continuous functions on the interval $[t,b]$. Let $Y(x)$ be a random variable with mean

$$E[Y(x)] = \beta_0 f_0(x) + \cdots + \beta_k f_k(x)$$

and variance

$$\text{var } Y(x) = \sigma^2(x) > 0 \quad \text{for } x \in [t,b].$$

Let $\hat{Y}(x)$ be the weighted, by reciprocal variance, least squares estimate of $E[Y(t)]$ based on size samples of n observations taken at points x_i in the subinterval $[a,b]$ of $[t,b]$. Since $g_j(x) = f_j(x)/\sigma(x), (j = 0, \ldots, k)$ is also a Chebyshev system on $[t,b]$, one may use the results of Hoel [1966] pertaining to that system to conclude that var[$\hat{Y}(t)$] can be minimized by using only $k + 1$ points in [a,b], provided the points and the number of observations at each point are suitably chosen. These points x_i are such that they make a regression function of the above form satisfy

$$E[Y(x_i)] = (-1)^i \sigma(x_i) \quad (i = 0, \ldots, k)$$

and are such that the graph of this regression function in the interval [a,b] will be inside the band formed by the two curves $y = \sigma(x)$ and $y = -\sigma(x)$. Figure 4 shows the geometry that characterizes the optimizing points for the special case of $k = 3$ for a regression function that satisfies the preceding conditions.

Since the optimizing points in [a,b] are such that the value of the regression function is either $\sigma(x)$ or $-\sigma(x)$ and such that its derivative is the same as the derivative of $\sigma(x)$ or $-\sigma(x)$ at each such point, except a and b, a set of equations based on these facts can be written down and solved for the x_i. The weights can be obtained using the theory of Hoel and Jennrich [1979].

Finding an Optimal Design for Estimating P(t)

Let the probability of an animal developing cancer due to a dose of level x of a carcinogenic material be given by

$$P(x) = 1 - \exp\left[-\sum_{j=0}^{k} \alpha_j x^j\right] \quad (\alpha_j \geq 0). \tag{1}$$

The above model is widely used in the field of cancer dose response. Hoel and Jennrich [1979] take the model to be

$$P(x) = 1 - \exp\left[-\sum_{j=0}^{k} \beta_j x^j\right] = 1 - \exp[-B(x)], \tag{2}$$

where β_j need not be nonnegative. Using the theory of Hoel [1966] pertaining to the Chebyshev's system, one can show that $k + 1$ dose levels are sufficient for asymptotic optimality. For both theoretical and practical considerations, one is better off using as few points as possible, and $k + 1$ is the smallest number that enables the model to be estimated.

Hoel and Jennrich [1979] show how to find a design that will minimize the asymptotic variance of an estimate of P(t) where t is any point in (0,a) and observations are taken in [a,b]. They do so by reducing the problem to that of optimal design for extrapolation in a Chebyshev regression model.

Let $x_0, \ldots, x_k$ be any set of $k + 1$ distinct points in [a,b]. Let $N_0, \ldots, N_k$ be the number of trials of an experiment that is performed at those points and let $R_0, \ldots, R_k$ denote the number of positive responses at the corresponding dose levels. Since there are $k + 1$ parameters and $k + 1$ sample points, estimates of

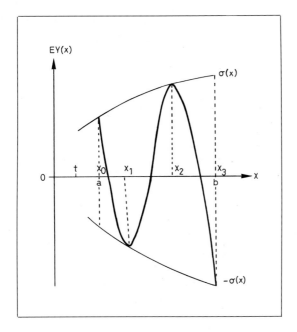

Fig. 4

the β values can be obtained either by the method of maximum likelihood or least squares by solving for β from the equations

$$\sum_{j=0}^{k} \beta_j x_j^i = -\ln(1 - \hat{p}_i) = \hat{z}_i \quad (i = 0, \ldots, k), \quad \text{where} \tag{3}$$

$$\hat{P}(x_i) = R_i/N_i = \hat{p}_i \quad (i = 0, \ldots, k).$$

Also,

$$\sqrt{N_i}(\hat{z}_i - z_i) \sim \sqrt{N_i}(\hat{p}_i - p_i)g'(p_i) = \sqrt{N_i}(\hat{p}_i - p_i)/(1 - p_i) \tag{4}$$

where $g(x) = -\ln(1 - x)$, $p_i = P(x_i)$ and $z_i = -\ln[1 - P(x_i)]$.
Let $N_i = c_i N$, $c_i \geq 0$ and $\Sigma c_i = 1$. Then $\hat{z}_0, \ldots, \hat{z}_k$ are independent and asymptotically normal variables. Hence, as $N \to \infty$,

$$\text{var}\{\sqrt{N_i}(\hat{z}_i - z_i)\} \sim \text{var}\frac{[\sqrt{N_i}(\hat{p}_i - p_i)]}{(1 - p_i)^2} = \frac{p_i q_i}{(1 - p_i)^2} = \frac{p_i}{q_i}. \tag{5}$$

Since the sample function $\hat{P}(x) = 1 - \exp[\hat{B}(x)]$ passes through the $k + 1$ sample points, its equation can be written in the form

$$\hat{P}(x) = 1 - \exp\left[-\sum_{j=0}^{k} L_j(x)\hat{z}_j\right] \tag{6}$$

where $L_j(x)$ is the Lagrange polynomial of degree k that satisfies $L_j(x_i) = \delta_{ij}$. Then, letting $g(x) = e^{-x}$, and proceeding as in (4), one obtains

$$\sqrt{N}[\hat{P}(x) - P(x)] \sim \Sigma L_j(x)Q(x)\sqrt{N}(\hat{z}_j - z_j),$$

where $Q(x) = 1 - P(x)$. If $v_j^2 = p_j/q_j$, then

$$\text{var}\{\sqrt{N}[\hat{P}(x) - P(x)]\} \sim \Sigma L_j^2(x)Q^2(x)(v_j^2/c_j).$$

If the c_j are chosen to minimize the asymptotic variance at $x = t$, then at that point

$$\text{var}\{\sqrt{N}[\hat{P}(t) - P(t)]\} \sim \left[\sum_j |L_j(t)Q(t)|v_j\right]^2. \tag{7}$$

So, the minimization of this variance is now reduced to minimizing

$$G(t) = \Sigma |L_j(t)Q(t)|v_j \tag{8}$$

over all choices of $x_0, \ldots, x_k$ where $a \leq x_0 < \ldots < x_k \leq b$.
Now applying the theory developed by Hoel and Jennrich [1979], (which assures the existence of an optimal set of points) one can obtain an optimal set of points by finding a regression function of the form

$$y(x) = \sum_{j=0}^{k} \beta_j x^j Q(x) \tag{9}$$

that satisfies the geometrical properties of figure 4 with $\sigma(x) = [P(x)Q(x)]^{1/2}$. The authors recommend the method of nonlinear least squares technique for finding optimal designs (if a computer is available) as an alternative approach.

Optimal Design for Estimating $P(t) - P(0)$. Hoel and Jennrich [1979] show that the same set of $k + 1$ points that are optimal for estimating $P(t)$ are also optimal for estimating $P(t) - P(0)$.

Optimal Design for Estimating t such that $P(t) = c$. Let $P(x) = F(x,\beta)$ where the β values occur in the exponent of $P(x)$. As an estimate of $\hat{t}$, choose t that satisfies $F(\hat{t},\hat{\beta}) = c$. By expanding F about t and β, one can show that asymptotically

$$\sqrt{N}(\hat{t} - t) \approx -\left(\frac{\partial F}{\partial x}\right)^{-1} \sqrt{N}[F(t,\hat{\beta}) - F(t,\beta)]. \tag{10}$$

Hence the asymptotic variance of t is a constant multiple of the asymptotic variance of $F(t,\hat{\beta}) = \hat{P}(t)$. This constant depends on t and β, but not on the design. A design that is asymptotically optimal for estimating $P(t)$ is also asymptotically

Optimal Designs in Carcinogen Experiments

optimal for estimating t. A corresponding result holds if one is interested in estimating t such that $P(t) - P(0) = c$.

Hoel and Jennrich [1979] point out that asymptotically it does not matter whether we consider restricted polynomials ($\Sigma \alpha_j x^j$) or unrestricted polynomials ($\Sigma \beta_j x^j$) in their model. They provide numerical techniques for obtaining optimal designs.

Examples

For obtaining optimal designs, one should know the response function $P(x)$. Preliminary experiments can be carried out in order to determine the prior probability function $P(x)$.

For illustrating the relative advantage of using optimal designs on models of the type considered here, the data of two experiments that were considered appropriate for this type of modelling will be studied. The first data is cited in Guess et al. [1977] and the second by Guess and Crump [1976].

Table 9.3. Data on vinyl chloride and benzopyrene

Vinyl chloride			Benzopyrene		
dose (x)	number of animals	number responding	dose (x)	number of animals	number responding
0	58	0	6	300	0
50	59	1	12	300	4
250	59	4	24	300	27
500	59	7	48	300	99

The maximum likelihood fit of $P(x)$ to the data of vinyl chloride in table 9.3 using model (1) as given by Guess et al. [1977] is

$P_1(x) = 1 - \exp(-0.000267377x)$.

The corresponding fit of $P(x)$ to the data of benzopyrene in table 9.4 using model (1) of Guess and Crump [1976] is

$P_2(x) = 1 - \exp(-0.000097x^2 - 0.0000017x^3)$.

Let $t = 0.5$ be the target value and let $[a,b] = [1,500]$. Then t falls outside of this interval as required by the theory of extrapolation [Hoel and Jennrich, 1979]. Since there will not be any appreciable difference in the optimal design, we consider $[a,b] = [0,500]$, thereby reducing the problem to that of interpolation.

Using $P_1(x)$ and carrying out the computations, the authors obtain, for estimating $P(t) - P(0)$,

$x_0 = 0$, $x_1 = 47.2$, $x_2 = 326$, $x_3 = 500$

$n_0 = 0$, $n_1 = 0.84N$, $n_2 = 0.12N$, $n_3 = 0.04N$

$\text{var}[\hat{P}(t) - \hat{P}(0)] = 3.36 \times 10^{-6}$,

where N denotes the total number of observations. (Note that $k = 3$ since 4 dose levels were used.) The corresponding variance for the spacing and weighting of table 9.4 was found to be 3.6 times as large as that for the optimal design.

General Comments. Empirical results carried out by the authors indicate that an optimal design is likely to yield better estimates of a response function value than will a traditional design. Optimal designs are insensitive to target values selected, provided it is small relative to the interval of experimentation. Also, optimal designs are somewhat insensitive to those modifications of the prior function which do not change the nature of the polynomial in the exponent of $P(x)$.

Since $P(x) - P(0)$ is approximately linear for small values of x, one might consider fitting a linear function and use it for estimation, realizing of course that it would be biased. Although the variance decreases rapidly as the degree of the polynomial in $P(x)$ decreases, the variance increases rapidly as the interval of observations is shortened. The gain in decreasing the bias by considering a short interval near the origin may be considerably setoff by the increase in the variance.

Asymptotic confidence intervals for $P(t)$ or $D(t) = P(t) - P(0)$ can be obtained by using the asymptotic normality of $\hat{P}(t)$ or $\hat{D}(t) = \hat{P}(t) - \hat{P}(0)$ and the use of formula (7) and

$$\text{var}\{\sqrt{N}[\hat{D}(x) - D(x)]\} \sim [\Sigma | L_j(0)Q(0) - L_j(x)Q(x) | v_j]^2. \tag{11}$$

Corresponding confidence intervals for t in the inverse problem can be obtained by using asymptotic normality and formulas (7) and (10) or (11) and (10).

10 Bayesian Bioassay

10.1. Introduction

In a quantal bioassay, there is a set of dose levels at which experimental units are tested. Each unit has a threshold or tolerance level. If the dose is less than the tolerance level, the subject does not respond; otherwise a response is observed. Let F(t) denote the tolerance distribution which is the proportion of the population with tolerance levels less than or equal to dose level t. At each dose level we observe a binomial variable with parameters n_i, the number of units tested at the ith dose and $p_i = F(t_i)$ where t_i is the ith dose. Often, there is some prior information available from previous assays using similar subjects and similar dose levels. We would like to incorporate such prior information into our estimation procedures.

Kraft and van Eeden [1964] characterize the class of all prior distributions for F, find the corresponding Bayes estimates for quadratic loss functions and apply the results of LeCam [1955] to show completeness of the closure of this class of estimates for a certain topology.

Let W(x) be an arbitrary (fixed) distribution function. If G, a nondecreasing bounded between 0 and 1 function on the real line, is the statistician's decision, and F is the true distribution determining the distribution of k (the number of responses out of n at dose level t) then the loss is

$$L(F,G) = \int (F - G)^2 dW. \qquad (1)$$

For this loss the Bayes estimator is the conditional expectation of the process for given k.

In a typical bioassay problem, $F = F_\theta = F_0(x - \theta)$ for some fixed F_0. Then, the usual loss function is

$$L(F_{\theta_1}, F_{\theta_2}) = (\theta_1 - \theta_2)^2$$

since

$$[F_{\theta_1}(u) - F_{\theta_2}(u)]^2 = (\theta_1 - \theta_2)^2 [F'_\xi(u)]^2 \quad (\theta_1 < \xi < \theta_2).$$

The loss function of Kraft and van Eeden [1964] is

$$\int (F_{\theta_1} - F_{\theta_2})^2 dW = (\theta_1 - \theta_2)^2 \int [F'_\xi(u)]^2 dW.$$

The two loss functions are in local agreement if

$$\int [F'_\xi(u)]^2 dW < M.$$

Based on the method called z-interpolation, Kraft and van Eeden [1964] were able to compute explicitly the estimates of F(t) for t of the form $j/2^l$, j is odd and $l > 3$.

The well-known nonparametric estimator for the mean of F(t) is the Spearman-Karber estimator which is the maximum likelihood estimator when we have equal spacing of doses and equal sample sizes at each dose. This is given by

$$\hat{\mu}_{S-K} = t_T + (d/2) - d \sum_{i=1}^{T} \hat{p}_i \qquad (2)$$

where d is the spacing between doses and t_T is the largest dose, and $\hat{p}_i = k_i/n_i$, k_i denoting the number of responses at dose i. Notice that this estimator makes no use of the available prior information.

Bayesian approach to the bioassay problem has been considered by Kraft and van Eeden [1964], Freeman [1970], Ramsey [1972], Tsutakawa [1972] and Wesley [1976]. The last author, besides proposing new estimators, has surveyed the previous work and has served as a basis for part of this chapter.

10.2. *The Dirichlet Prior*

An important and useful prior in nonparametric Bayes approach is the Dirichlet process prior. A good description of this prior and its properties are given by Ferguson [1973].

Definition

Let Θ be a set and β a σ-field of subsets of Θ. Let m be a finite, nonnull, nonnegative, finitely additive measure on (Θ, β). A random probability measure P on (Θ, β) is said to be a Dirichlet process on (Θ, β) with parameter m if for every $K = 1, 2, \ldots$ and measurable partition $B_1, B_2, \ldots, B_K$ of Θ, the joint distribution of the random probabilities $[P(B_1), \ldots, P(B_K)]$ is a Dirichlet process with parameters $[m(B_1), \ldots, m(B_K)]$ [Antoniak, 1974].

We can think of m as a measure generated by a multiple of a distribution function

$$m((-\infty, t]) = M \cdot \alpha(t)$$

where M is a positive real number and α is a distribution function.

For any integer $T \geq 1$ and any set of real numbers $t_1, \ldots, t_T$ such that $t_1 \leq \ldots \leq t_T$, let

$Y_1 = P((-\infty, t_1]|F) = F(t_1)$

$Y_2 = P((t_1, t_2]|F) = F(t_2) - F(t_1)$

$\vdots$

$Y_{T+1} = P((t_T, \infty)|F) = 1 - F(t_T)$

and

$\beta_1 = m((-\infty, t_1]) = M \cdot \alpha(t_1)$

$\beta_2 = m((t_1, t_2]) = M \cdot [\alpha(t_2) - \alpha(t_1)]$

$\vdots$

$\beta_{T+1} = m((t_T, \infty)) = M[1 - \alpha(t_T)]$.

Then

$Y = (Y_1, \ldots, Y_{T+1})' \sim \text{Dirichlet}(\beta_1, \ldots, \beta_{T+1}) = D_T(\beta_1, \ldots, \beta_{T+1})$.

Notice that the family of prior distributions D_T is a consistent family. That is, all marginal distributions belong to the same class of priors. For instance, the marginal distribution of $Y_1, \ldots, Y_T$ is $D_{T-1}(\beta_1, \ldots, \beta_{T-1}, \beta_T + \beta_{T+1})$. Thus, the tolerance distribution F is defined as a random variable.

The density of $(Y_1, \ldots, Y_{T+1})$ can be written as

$$f_Y(y_1, \ldots, y_T) = \frac{\Gamma(M)}{\prod_{i=1}^{T+1} \Gamma(\beta_i)} \prod_{j=1}^{T+1} y_j^{\beta_j - 1}, \quad \text{where}$$

$$y_{T+1} = 1 - \sum_{i=1}^{T} y_i, \quad \text{on} \quad S = \left\{(y_1, \ldots, y_T): y_1 \geq 0, \ldots, y_T \geq 0 \sum_{i=1}^{T} y_i \leq 1\right\}$$

and zero elsewhere. Alternatively, we can write the density of $(Y_1, \ldots, Y_{T+1})$ as proportional to

$$\prod_{j=1}^{T+1} y_j^{\beta_j} \tag{3}$$

where averaging is done with respect to the measure

$$dv = \prod_{j=1}^{T} dy_j \bigg/ \prod_{j=1}^{T+1} y_j \tag{4}$$

one can interpret α as a 'prior' distribution function summarizing one's prior idea of the tolerance distribution function. It is the mean of the random distribution function F. M is equivalent to the amount of confidence we have in the prior, or the number of observations that the prior distribution function is worth.

10.3. The Bayes Solution for Squared Error Loss

With the notation developed in sections 10.1 and 10.2, the Bayes estimate of F(t) under squared error loss is the mean of the posterior distribution of F(t). According to Antoniak [1974], for more than one dose, the posterior distribution can be written as a mixture of Dirichlet distributions, and hence the mean can easily be calculated. Then the mean of the tolerance distribution can be estimated by the mean of the estimated tolerance distribution $\hat{F}(t)$: If $\alpha(t)$ is normal (μ_p, σ^2) then the mean of the estimated tolerance distribution is

$$\mu_{\hat{F}} = \mu_p + \sigma \left\{ \sum_{i=0}^{T} \frac{\hat{F}(t_{i+1}) - \hat{F}(t_i)}{\Phi\left(\frac{t_{i+1} - \mu_p}{\sigma}\right) - \Phi\left(\frac{t_i - \mu_p}{\sigma}\right)} \left[\varphi\left(\frac{t_i - \mu_p}{\sigma}\right) - \varphi\left(\frac{t_{i+1} - \mu_p}{\sigma}\right) \right] \right\} \quad (5)$$

where $\hat{F}(t_i)$ denotes the estimate of $F(t_i)$ at dose t_i and we take $t_0 = -\infty$ and $t_{T+1} = \infty$. This estimate is cumbersome to compute since it involves nested summation signs. For a proof of (5), see Wesley [1976, Appendix C].

10.4. The Alternate Bayes Approaches

Ramsey's Approach

The posterior distribution of F is proportional to the product of the prior density and the likelihood. The joint mode of the posterior will be used to summarize the posterior distribution. Note that the log of the posterior density is proportional to

$$\sum_{i=1}^{T+1} k_i \ln F(t_i) + \Sigma(n_i - k_i) \ln[1 - F(t_i)] + \sum_{i=1}^{T+1} M\delta_i \ln[F(t_i) - F(t_{i-1})],$$

where $\delta_i = \alpha(t_{i+1}) - \alpha(t_i), i = 1, \ldots, T+1, \sum_i^{T+1} \delta_i = 1$ and M is a positive constant.

First, we reparametrize the posterior density by setting $\theta_i = F(t_i) - F(t_{i-1})$ for $i = 1, \ldots, T$. Then $F(t_i) = \theta_1 + \cdots + \theta_i$ and the θ_i are subjected to the constraints $\theta_i + \cdots + \theta_T = 1$. Now, we maximize the natural logarithm of the posterior density subject to the constraint using the Lagrange method. The difference between the partial derivatives with respect to θ_i and θ_{i+1} yields (6).

Setting $\dfrac{\partial}{\partial F(t_i)}(\ln \text{posterior}) = 0$ we obtain

$$\frac{n_i}{\hat{F}(t_i)[1 - \hat{F}(t_i)]} \left[\frac{k_i}{n_i} - \hat{F}(t_i) \right] = M \left[\frac{\delta_{i+1}}{\hat{F}(t_{i+1}) - \hat{F}(t_i)} - \frac{\delta_i}{\hat{F}(t_i) - \hat{F}(t_{i-1})} \right] \quad (6)$$

if t_i is an observational dose and

$$0 = \frac{\alpha_{i+1} - \alpha(t)}{\hat{F}(t_{i+1}) - \hat{F}(t_i)} - \frac{\alpha(t) - \alpha_i}{\hat{F}(t_i) - \hat{F}(t_{i-1})}, \alpha_i = \alpha(t_i) \quad (7)$$

if t is not an observational dose and $t_i < t < t_{i+1}$. Notice that we were able to obtain (7) because at t the prior is still defined and for $t_i < t < t_{i+1}$, we replace the interval $(t_i; t_{i+1})$ by the two intervals $(t_i; t)$ and $(t; t_{i+1})$. However, the likelihood does not have the term associated with t.

When $M = \infty$, the posterior distribution is dominated by the prior distribution. The prior d.f. tends to a degenerate d.f. giving probability one to the prior mode. On the other hand when $M \to 0$, the mode of the joint density is the isotonic regression estimator introduced by Ayer et al. [1955]. The solution may be written as

$$\hat{F}(t) = \begin{cases} 0 & \text{for } t < t_1 \\ \min_{i \leqslant s \leqslant T} \max_{1 \leqslant r \leqslant i} \left[\sum_{j=r}^{s} k_j \bigg/ \sum_{j=r}^{s} n_j \right] & \text{for } t_i \leqslant t \leqslant t_{i+1} \text{ and } i = 1, \ldots, T, \end{cases} \quad (8)$$

where $t_1, \ldots, t_T$ are the observational doses and t_{T+1} is taken to be ∞. The modal function given by (8) is uniquely defined only at the observational doses $t_1, \ldots, t_T$. The interpolation method is arbitrary, subject to the constraint of monotonicity. For $0 < M < \infty$, the joint posterior density is convex and unimodal. A unique nondecreasing function $\hat{F}(t)$ referred to as the posterior modal function, exists such that for any set of doses $(t_1, \ldots, t_T)$ the mode of the posterior density occurs at $[\hat{F}(t_1), \ldots, \hat{F}(t_T)]$.

Setting the bracketed quantities on the left and right side of (6) equal to zero would define $\hat{F}(t)$ as the mode of the likelihood function and the mode of the prior density, respectively. Since the bracketed quantity on the left side of (6) is weighted by the Fisher information in the likelihood about $\hat{F}(t_i)$, one can interpret M as a measure of the information in the prior. To determine $\hat{F}(t)$ first solve the T-simultaneous equations in (6) for the $\hat{F}(t_i)$. Equation (7) tells how one should interpolate or extrapolate.

Suppose we now wish to determine the dose level $\hat{t}$ such that $\hat{F}(\hat{t}) = \gamma$. If for some t_i, $\hat{F}(t_i) = \gamma$, then $\hat{t} = t_i$. If not, determine the pair of observed doses t_i and t_{i+1} for which $\hat{F}(t_i) < \gamma < \hat{F}(t_{i+1})$ (with the understanding that $t_0 = -\infty$ and $t_{T+1} = \infty$). The potency at $\hat{t}$ between t_i and t_{i+1} may be included in the prior since this has no effect on the posterior at the observational doses; hence we include it in the posterior. Setting the partial derivative w.r.t. $F(\hat{t})$ of the joint posterior density to zero yields:

$$\frac{\alpha_{i+1} - \hat{\alpha}}{\hat{F}(t_{i+1}) - \hat{F}(\hat{t})} = \frac{\hat{\alpha} - \alpha_i}{\hat{F}(\hat{t}) - \hat{F}(t_i)} \quad \text{where } \hat{\alpha} = \alpha(\hat{t}).$$

That is

$$\frac{\alpha_{i+1} - \hat{\alpha}}{\hat{F}(t_{i+1}) - \gamma} = \frac{\hat{\alpha} - \alpha_i}{\gamma - \hat{F}(t_i)}.$$

Letting

$$\xi = [\gamma - \hat{F}(t_i)]/[\hat{F}(t_{i+1}) - \hat{F}(t_i)] \quad (9)$$

we obtain

$$\hat{\alpha} = \alpha_{i+1}\xi + \alpha_i(1 - \xi) = \alpha_i + \xi(\alpha_{i+1} - \alpha_i). \quad (10)$$

The interpolation formulae (9) and (10) are linear and they impose the condition that the posterior mode $\hat{F}(t)$ should have the same shape as the prior mode $\alpha(t)$ piece-wise between observational doses.

Ramsey [1972] provides several examples based on artificial data. In all these examples the posterior mode is found by maximizing the posterior density subject to inequality constraints rather than solving (6) directly. The author employs an algorithm which converts the problem into an unconstrained maximization problem by the introduction of a 'penalty' function. The examples suggest that one observation per dose may be the best design for estimating ED_{50}.

Turnbull's Approach

Turnbull [1974, 1976] describes a method for estimating (nonparametrically) a distribution F when the data are incomplete due to censoring. In the bioassay, the data is either right or left censored. The technique of Turnbull [1974, 1976] consists of maximizing the likelihood by an iterative procedure called the 'self-consistency algorithm.' To adapt his procedure to the Bayesian situation, we treat the prior distribution as 'prior' observations. Let $T \equiv 3$, and consider the Dirichlet prior

$$f_Y = c y_1^{\beta_1} y_2^{\beta_2} y_3^{\beta_3} y_4^{\beta_4}$$

where $y_1 = F(t_1)$, $y_2 = F(t_2) - F(t_1)$, $y_3 = F(t_3) - F(t_2)$ and $y_4 = 1 - F(t_3)$, and c is a constant. We set $\beta_i = 1$ for $i = 1, \ldots, 4$ so that M, the prior sample size, is 4. Suppose that we have a response at t_1, a non-response at t_2 and a response at t_3. Then the usual likelihood is

$$f_{K|Y} = c' y_1^{k_1}(1-y_1)^{1-k_1}(y_1+y_2)^{k_2}(1-y_1-y_2)^{1-k_2}(y_1+y_2+y_3)^{k_3}y_4^{1-k_3}$$
$$= c' y_1 (1 - y_1 - y_2)(y_1 + y_2 + y_3).$$

On the other hand, if we treat the prior as additional observations, the likelihood becomes

$$f_K^* = c'' y_1 y_2 y_3 y_4 y_1 (1 - y_1 - y_2)(y_1 + y_2 + y_3),$$

which is proportional to the posterior distribution

$$f_Y f_{K|Y} \Big/ \int_S f_Y f_{K|Y} dY \tag{11}$$

under prior f_Y and observations K. Thus the self-consistency algorithm with the prior as additional observations maximizes the posterior as we hoped. In general let $\beta_i = \gamma_i/g$ where γ_i and g are integers. Then the posterior is proportional to

$$f_1 = \left(\prod_{i=1}^{T+1} y_i^{\beta_i}\right)\left(\prod_{j=1}^{T} z_j^{k_j}(1-z_j)^{n_j-k_j}\right) = \left[\prod_{i=1}^{T+1} y_i^{\gamma_i} \prod_{j=1}^{T} z_j^{k_j g}(1-z_j)^{(n_j-k_j)g}\right]^{1/g},$$

where

$$z_j = \sum_{i=1}^{j} y_i.$$

If we assume that there are γ_i prior observations in the interval y_i, then multiplying each actual observation by the common denominator g, the self-consistency algorithm will maximize $f = f_1^g$. Since the transformation is monotone, the maximizing values of $\{y_i\}$ will also maximize f_1 which is proportional to the posterior.

Suppose we have doses t_1, t_2, t_3 and we desire an estimate at dose t^* where $t_1 \leqslant t^* \leqslant t_2$. Then we set up the new intervals $(-\infty, t_1)$, (t_1, t^*), (t^*, t_2), (t_2, t_3), (t_3, ∞). The number of prior observations assigned to the new intervals (t_1, t^*) and (t^*, t_2) will be according to the prior distribution function and their sum will equal the number previously assigned to (t_1, t_2). The estimates of the tolerance distribution at a nonobservational dose t are more easily obtained from (7) yielding

$$\hat{F}(t) = \hat{F}(t_i) + \frac{\alpha(t) - \alpha(t_i)}{\alpha(t_{i+1}) - \alpha(t_i)} [\hat{F}(t_{i+1}) - \hat{F}(t_i)] \quad \text{for } t_i < t \leqslant t_{i+1} \quad (i = 0, \ldots, T) \tag{12}$$

where we set $t_0 = -\infty$ and $t_{T+1} = \infty$. The expression for the mean of F(t) is again given by (5).

The estimates based on the mode of the posterior are easier to compute than those based on quadratic loss and seem to give estimates very close to those obtained from the mean of the posterior, especially for symmetric distributions.

10.5. *Bayes Binomial Estimators*

Let us consider a simple dose experiment consisting of n subjects. The conjugate prior is the beta distribution $B[M\alpha(t), M(1 - \alpha(t))]$. The Bayes estimate of p, the probability of a response at t, is

$$\hat{p}(BB) = [M\alpha(t) + k]/(M + n) \tag{13}$$

where k is the number of responses out of n trials.

If we follow this procedure for each dose level, we would have estimates of the tolerance d.f. at each dose and these can be combined to obtain a Spearman-Karber type of an estimate for the mean. This will be called the Bayes-binomial estimator introduced by Wesley [1976] and will be denoted by $\hat{\mu}_{BB}$. For equally spaced dose levels (d being the common distance) and equal sample sizes n at each dose, we have

$$\hat{\mu}_{BB} = t_T + (d/2) - d \sum_{i=1}^{T} \hat{p}_i(BB) = t_T + \frac{d}{2} - \frac{dM}{M+n} \Sigma \alpha_i - \frac{dn}{M+n} \Sigma \frac{k_i}{n}$$

$$= \hat{\mu}_{S-K} + \frac{dM}{M+n} \Sigma (\hat{p}_i - \alpha_i).$$

$\hat{\mu}_{BB}$ will be close to $\hat{\mu}_{S-K}$ if the observed proportion of responses at each dose is close to the prior d.f. Although this estimating procedure treats each dose as an independent binomial experiment, it is relatively easy to compute.

10.6. *Selecting the Prior Sample Size*

In the nonparametric Bayes framework, the prior sample size M can be viewed as the number of 'observations' we are willing to give to the expected proportion of responses compared to the number of actual observations we are going to take.

In the parametric Bayes approach, let F be normal with mean μ and variance τ^2 and let μ be normal with mean 0 and variance σ^2. Since we will be concerned with the ratio σ/τ, without loss of generality we can set $\tau = 1$. Hence,

$$E[F(t)] = \sigma^{-1} \int_{-\infty}^{\infty} \Phi(t - \mu)\phi(\mu/\sigma)d\mu$$

$$E[F^2(t)] = \sigma^{-1} \int_{-\infty}^{\infty} \Phi^2(t - \mu)\phi(\mu/\sigma)d\mu.$$

Then let

$$\text{Var}^* F(t) = E[F^2(t)] - \{E[F(t)]\}^2.$$

On the other hand, if we approach the problem using the Dirichlet prior, we would take our prior d.f. $\alpha(t)$ to be $\Phi(t)$ (since $\tau = 1$). α is the mean of the Dirichlet prior and should correspond to the underlying tolerance distribution function F(t).

$$\text{Var}' F(t) = \alpha(t)[1 - \alpha(t)]/(M + 1)$$

where Var' denotes the variance under the Dirichlet prior. Equating $\text{Var}^* F(t)$ with $\text{Var}' F(t)$ and solving for M, we have

$$M(t) = \frac{\alpha(t)[1 - \alpha(t)]}{\text{Var}^* F(t)} - 1.$$

We hope that this estimate of M will be fairly constant over the dose range; it also depends on σ/τ. We could consider averaging over the doses t_j or consider a conservative estimate of M to be

$$M = \min_{1 \leq j \leq T} M(t_j).$$

Example

For $\sigma = 1$ (or $\sigma = \tau$), $\text{Var}^* F(0) = 1/12$. If $\alpha(t) = \Phi(t)$, then $\alpha(0) = 1/2$. Thus $M(0) = 2$.

10.7. *Adaptive Estimators*

Wesley [1976] has modified the cross-validation approach of Stone [1974] and the pseudo-Bayes estimators of Bishop et al. [1975] to the bioassay problem.

We will consider these in the following. At each dose we have a separate binomial experiment. Hence, we can apply the multinomial techniques at each dose level to estimate the probability of a response.

Alternatively, we can estimate the multinomial cell frequencies by $N(\tilde{p}_j - \tilde{p}_{j-1})$ where N is the total number of experimental units and $\tilde{p}_j$ is the isotonic regression estimate of the tolerance distribution at dose j. Then, the estimate of the mean of the tolerance distribution is of the Spearman-Karber type and is given by

$$\hat{\mu} = t_T + (d/2) - d \sum_{i=1}^{T} \left(\sum_{j=1}^{i} \tilde{q}_j \right) = t_T + (d/2) - d \sum_{i=1}^{T} (T - i + 1)\tilde{q}_i$$

where $\tilde{q}_j$ is the estimate of the multinomial cell probability of having a tolerance greater than dose $j - 1$ and not exceeding dose j.

Using the multinomial formulas of Stone [1974] with x_i estimated from the differences between successive isotonic regression estimates gives

$$w = \left[(T \cdot n)^2 - \sum_{j=1}^{T+1} \tilde{x}_j \right] \left\{ T \cdot n - 2 \sum_{j=1}^{T+1} \tilde{x}_j [\tilde{x}_j - \lambda_j (T \cdot n - 1)] + T \cdot n \sum_{j=1}^{T+1} [\tilde{x}_j - \lambda_j (T \cdot n - 1)]^2 \right\}^{-1} \quad (14)$$

where

x_j = number of observations in the j-th cell,

λ_j = prior probability for the j-th cell,

$\tilde{x}_j = T \cdot n (\tilde{p}_j - \tilde{p}_{j-1})$

assuming equal sample sizes at each dose. Then

$\hat{q}_j(w) = w\lambda_j + (1 - w)(\tilde{p}_j - \tilde{p}_{j-1})$ and

$$\hat{\mu}_{Stone} = t_T + (d/2) - d \sum_{i=1}^{T} (T - i + 1)\hat{q}_i(w). \quad (15)$$

If $w = M/(M + N)$, then we get a form similar to that of the Bayes-binomial estimator. This cross-validation approach can be viewed as choosing a value of M based on the data.

Wesley [1976] obtains another adaptive estimator by applying the pseudo-Bayes approach [Bishop et al., 1975] to the Bayes-binomial estimate.
Recall

$$\hat{p}_j(BB) = \frac{M \cdot \alpha(t_j)}{M + n} + \frac{n}{M + n} \frac{k_j}{n} \quad \text{and}$$

$$\hat{\mu}_{BB} = t_T + (d/2) - \frac{dM}{M + n} \Sigma \alpha_i - \frac{dn}{M + n} \Sigma \frac{k_i}{n}.$$

So

$$MSE(\hat{\mu}_{BB}) = \left[\frac{dM}{M + n} \Sigma(p_i - \alpha_i) + t_T + \frac{d}{2} - d\Sigma p_i - \mu \right]^2 + \frac{n^2}{(M + n)^2} \frac{d^2}{n} \Sigma p_i (1 - p_i).$$

Maximizing this for M gives

$$M = \frac{d\Sigma p_i(1-p_i) - n\left[t_T + \dfrac{d}{2} - d\Sigma p_i - \mu\right][\Sigma(p_i - \alpha_i)]}{d[\Sigma(p_i - \alpha_i)]^2 + \left(t_T + \dfrac{d}{2} - d\Sigma p_i - \mu\right)(\Sigma(p_i - \alpha_i))}.$$

If we estimate μ by $\hat{\mu}_{S-K}$ and p_i by $\hat{p}_i = k_i/n$, then

$$\hat{M}_{BB} = \frac{\sum_1^T(\hat{p}_i - \hat{p}_i^2)}{\left[\sum_1^T(\hat{p}_i - \alpha_i)\right]^2} = \frac{n\sum_1^T k_i - \sum_1^T k_i^2}{\left(\sum_1^T k_i\right)^2 - 2n\sum_1^T k_i \sum_1^T \alpha_i + n^2\left(\sum_1^T \alpha_i\right)^2}.$$

Hence, Wesley's estimate of the probability of response at t_j is

$$\hat{p}_j(\hat{M}_{BB}) = \frac{\hat{M}_{BB} \cdot \alpha_j}{\hat{M}_{BB} + n} + \frac{n}{\hat{M}_{BB} + n}\frac{k_j}{n}, \text{ and}$$

$$\hat{\mu}(\hat{M}_{BB}) = t_T + \frac{d}{2} - d\Sigma \hat{p}_j(\hat{M}_{BB}). \tag{16}$$

10.8. *Mean Square Error Comparisons*

Let us assume 3 dose levels and 5 subjects at each dose, and the doses are equally spaced $d = 2$ units apart and are symmetric about the prior mean $= 0.6$. Also assume that the tolerance distribution is normal (μ, σ^2), where $\sigma^2 = 0.25$. Wesley [1976] graphs the mean square errors of various estimators. Note that the graph of MSE $(\hat{\mu}, \mu)$ for different values of μ is symmetric about the prior mean. For the Turnbull-Bayes and Bayes-binomial estimators, M is taken to be 1.0. Wesley [1976] draws the following conclusions:
Changing M has no effect on the MSE of the Spearman-Karber or the adaptive estimators. For large M, the MSE of Bayes-binomial approaches that of Turnbull-Bayes estimator. For small M, the Bayes-binomial is equivalent to the Spearman-Karber estimate.
When M is large, the Turnbull-Bayes and Bayes-binomial estimators have lower MSE at values of the true mean close to the prior mean (within 0.5–1 SD). However, they do increasingly worse as the true mean moves away from the prior mean. The Bayes-binomial estimator is not much worse than the Turnbull-Bayes estimator when the prior is accurate, and is better for true means far from the prior since it is less dependent on the means.
Wesley [1976] graphs the mean square errors for distributions other than normal, namely uniform, Cauchy, and angular. The proposed estimators are fairly robust w.r.t. the specific form of the tolerance distributions.

In conclusion, the experimenter must decide the relative merits of protection when the prior is wrong and accuracy when the prior is correct. If no prior knowledge is available use the Spearman-Karber estimate or go to a two-stage assay. If we have a good prior knowledge, the Turnbull-Bayes estimator is suitable. If the prior is accurate within 0.5–1 SD, setting M = 5 or 10 does a much better job than disregarding the prior information and simply using the Spearman-Karber estimator.

10.9. *Bayes Estimate of the Median Effective Dose*

Although the up and down and the stochastic approximation methods are simple to apply and have high efficiency, the sequence of dose levels they propose may be suboptimal, especially for small sample sizes. Furthermore, in the case of the up and down method, the choice of the stopping rule is left to the experimenter, so that the payoff between the cost of further experimentation and that of less accurate estimation is not considered explicitly. Marks [1962] considered the problem of Bayesian design in estimating (sequentially) the mean of a quantal probit response curve and obtained detailed results in the special cases when the prior for the median γ is a two-point distribution or when the dose-response curve is a step function. Freeman [1970] considered the sequential design of experiments for estimating the median lethal dose parameter of a quantal logistic dose-response curve. He employs the Bayesian decision theory to obtain a stopping rule and a terminal decision rule for minimizing the prior expectation of the total cost of observation plus (quadratic) estimation loss. He numerically evaluates the optimal strategies for the special cases in which observations can be obtained only at one, two or three dose levels. He compared the expected losses with those of the up and down method using a fixed sample size equal to the prior expectation of the number of trials under the sequential design. Surprisingly, he found that the efficiency of the up and down method is in excess of 90%. Freeman [1970] assumes that the scale parameter in the logistic response function is known. He employs the dynamic programming equations.
Recall that the logistic response function is given by

$$p = [1 + e^{-\beta(x-\gamma)}]^{-1} \quad (-\infty < \gamma < \infty)$$

where γ denotes the median and β the scale parameter. We observe a quantal response variable Y where

$$P(Y = 1) = p \quad \text{and} \quad P(Y = 0) = 1 - p.$$

Let r be the number of 1's in n independent trials. Then r is distributed binomially with parameters n and p. γ is assumed to have the conjugate prior given by

$$f(\gamma | r_0, n_0) \propto \frac{e^{r_0 \beta(x-\gamma)}}{[1 + e^{\beta(x-\gamma)}]^{n_0}}$$

which corresponds to the usual beta prior $\propto p_0^{r_0-1}(1-p_0)^{n_0-r_0-1}$ for p. Freeman [1970] takes $r_0 = 1$ and $n_0 = 2$ to make the relevant distributions proper. Let c denote the cost per observation (assumed to be independent of the outcome) and $k(\hat{\gamma} - \gamma)^2$ denote the loss incurred in estimating γ by $\hat{\gamma}$. This corresponds to the loss function

$$\frac{k}{\beta^2}\left\{\log\left[\frac{p(1-\hat{p})}{\hat{p}(1-p)}\right]\right\}^2$$

for p. At a general point (r,n), following Lindley and Barnett [1965], let

D(r,n) = loss incurred by stopping and using the optimal estimate for γ,

B*(r,n) = the loss incurred by taking one further observation and using the optimal strategy thereafter,

B(r,n) = the loss incurred by using the optimal strategy.

The dynamic programming equations are

$B^*(r,n) = c + (r/n)B(r+1, n+1) + [(n-r)/n]B(r, n+1)$,

$B(r,n) = \min[B^*(r,n), D(r,n)]$,

$D(r,n)$ = posterior variance of γ.

Computations yield

$$f(\gamma|r,x) = \frac{\beta \cdot (n + n_0 - 1)!}{(n + n_0 - r - r_0 - 1)!(r + r_0 - 1)!} \cdot \frac{e^{-(r+r_0)\beta(x-\gamma)}}{[1 + e^{-\beta(x-\gamma)}]^{n+n_0}}.$$

In order to find the posterior mean and variance of γ, one can make the transformation

$$z^{-1} = 1 + e^{-\beta(x-\gamma)}$$

and reduce the posterior density of γ to a beta density. Upon noting that

$$\frac{\partial^k}{\partial \alpha^k}[\Gamma(\alpha)\Gamma(\beta)/\Gamma(\alpha+\beta)] = \int_0^1 [\ln(1-z)]^k z^{\beta-1}(1-z)^{\alpha-1} dz$$

$$\frac{\partial^k}{\partial \beta^k}[\Gamma(\alpha)\Gamma(\beta)/\Gamma(\alpha+\beta)] = \int_0^1 (\ln z)^k z^{\beta-1}(1-z)^{\alpha-1} dz$$

$$\frac{\partial^2}{\partial \alpha \partial \beta}[\Gamma(\alpha)\Gamma(\beta)/\Gamma(\alpha+\beta)] = \int_0^1 (\ln z)[\ln(1-z)] z^{\beta-1}(1-z)^{\alpha-1} dz$$

one can easily obtain (for the choice of $r_0 = 1$, $n_0 = 2$)

$$E(\gamma|r,x) = x + \frac{1}{\beta}\left[\sum_{i=n-r+1}^{\infty} \frac{1}{i} - \sum_{i=r+1}^{\infty} \frac{1}{i}\right]$$

$$\text{Var}(\gamma|r,x) = \frac{1}{\beta^2}\left[\sum_{i=r+1}^{\infty} i^{-2} - \sum_{i=n-r+1}^{\infty} i^{-2}\right]$$

$D(r,n) = k \text{Var}(\gamma|r,x)$.

11 Radioimmunoassays

11.1. Introduction

Radioimmunoassays had their origin more than a decade ago, when antibodies capable of binding ^{131}I-labeled insulin were demonstrated in human subjects treated with a mixture of commercial beef and pork insulin. It was shown in the literature that the percentage of insulin bound to antibody decreased as the insulin concentration in the incubation mixtures was increased and unlabeled insulin could displace labeled insulin from the insulin-antibody complexes. Yallow and Berson [1970] provided an excellent survey of the technical aspects of radioimmunoassays which, unlike the traditional bioassays, are dependent on specific chemical reactions that obey the law of mass action and are not subject to errors introduced by the biological variability of test systems. McHugh and Meinert [1970] provided a theoretical model for statistical inference in isotope displacement immunoassay.

11.2. Isotope Displacement Immunoassay

McHugh and Meinert [1970] have developed a theoretical model for the bioassay of human insulin in an isotape displacement immunoassay, which include the following steps:
(1) the biochemical system as a basis for the model,
(2) statistical considerations including nonlinear curve fitting and the derivation of formulas for confidence limits,
(3) comparison of the theoretical model to the logit model,
(4) estimating inversely the unknown ('test') antigen potency from the results of steps (1) and (2).
A brief outline of the biochemical model is essential in understanding the theoretical model proposed by Meinert and McHugh [1968]. The immune system of the body produces antigens to fight the antibodies which threaten it. These antigens bind to the antibodies, rendering them harmless. The reaction can be formulated as

$$Ag + Ab \rightleftarrows AgAb \tag{1}$$

where Ag denotes antigen, Ab denotes antibody and $\rightleftarrows$ denotes a two-way reaction.

In developing an immunoassay, it has been found that the concentration of bound antigen, AgAb, is a curvilinear function of the initial concentration of antigen and antibody in the system. Thus, by measuring the AgAb in the system, one can infer the initial concentrations of either the antigen or antibody. However, measuring the bound antigen antibody complex may prove difficult. For this reason, part of the antigen in the system is labelled with radioactive traces such as iodine-131. The known quantities in the system are labelled antigen and the antibody. The presence of an unknown amount of unlabelled antigen in the system decreases the labelled antigen that is bound to the antibody and increases the proportion of the unbound labelled antigen. Measurements of the unbound antigen, or the radioactive antigen antibody complex, then allows for an indirect measure of the unlabelled antigen, the amount of which is unknown. Meinert and McHugh [1968] explain how a standard curve is developed, with which specimens of unknown potency can be compared. This standard curve is based on biochemical theory and the law of mass action. At equilibrium, the reaction in (1) continues with the reaction going to the right at a velocity v_F, and the reaction going to the left at velocity v_R. The rate of these velocities is denoted by k ($k = v_F/v_R$). By the law of mass action

$$k = [AgAb]/[Ag][Ab] \tag{2}$$

where [ ] denotes concentration. Then k can be written in terms of the initial concentration of antigen and antibody denoted by $[Ag_0]$ and $[Ab_0]$. Thus,

$$k = \frac{[AgAb]}{([Ag_0] - [AgAb])([Ab_0] - [AgAb])} \tag{3}$$

where

$[Ag_0] - [AgAb] = [Ag]$ and $[Ab_0] - [AgAb] = [Ab]$.

For ease of computations, let

$x' = [Ag_0]$, $\theta_1 = [Ab_0]$, $\eta' = [AgAb]$ and $\theta_2 = k$.

Then (3) becomes

$$\theta_2 = \eta'/(x' - \eta')(\theta_1 - \eta') \tag{4}$$

where the following conditions are dictated by the physical properties of the model.

$0 < \eta' < x'$, $\theta_1 > 0$, $\eta' < \theta_1$ and $\theta_2 > 0$.

Solving for η' gives

$$\eta' = \{(\theta_1 + x' - \theta_2^{-1}) - [(\theta_1 + x' - \theta_2^{-1})^2 - 4x'\theta_1]^{1/2}\}/2 \tag{5}$$

Note that here we took the negative root of the equation because the positive root will violate the assumption that $\eta' \leq \theta_1$. Let us write

$$x' = x + A_L$$

where A_L is the isotope-labelled antigen and x is the unlabelled antigen, and let

$$\eta' = \eta(A_L + x).$$

Then (5) can be rewritten as

$$\eta = \{(\theta_1 - \theta_2^{-1} + A_L + x) - [(\theta_1 - \theta_2^{-1} + A_L + x)^2 - 4\theta_1(A_L + x)]^{1/2}\}/2(A_L + x) \qquad (6)$$

which is the equation for the theoretical calibration curve for the deterministic model. Now let

$$\eta_i = Ey_{ij}$$

which is the true proportion of radioactive counts associated with bound antigen in the j-th tube in the i-th set of tubes.

Thus, due to errors in the measurement process, we have the stochastic model

$$y_{ij} = \text{expression (6) with x replaced by } x_i + \varepsilon_{ij} \quad (j = 1,\ldots,n_i \text{ and } i = 1,\ldots,s). \qquad (7)$$

The inverse of this calibration (standard) curve will give a value for x_i which is the concentration of unlabelled antigen in the system. Thus

$$x_i = [\theta_1 \theta_2 - \eta_i(1 - \eta_i)^{-1}]/\theta_2 \eta_i - A_L. \qquad (8)$$

Then the estimate of x_i, namely $\hat{x}_i$ is obtained by substituting the estimates for θ_1 and θ_2 in (8). However, estimation of θ_1 and θ_2 is somewhat difficult because equation (6) is nonlinear in these parameters. McHugh and Meinert [1970] make use of the Gauss-Newton iterative method in order to numerically approximate θ_1 and θ_2. The rates of convergence is rapid provided one uses the initial values proposed by the authors:

$$\hat{\theta}'_{1,0} = n^{-1}\Sigma z_{ij} - n^{-1}\Sigma w_{ij}\hat{\theta}_{2,0}$$

$$\hat{\theta}_{2,0} = [n\Sigma w_{ij}z_{ij} - (\Sigma w_{ij})(\Sigma z_{ij})][n\Sigma w_{ij}^2 - (\Sigma w_{ij})^2]^{-1} \quad \text{and}$$

$$\hat{\theta}_{1,0} = \hat{\theta}'_{1,0}/\hat{\theta}_{2,0}, \quad \text{where}$$

$$z_{ij} = y_{ij}/(1 - y_{ij}), \quad w_{ij} = y_{ij}x'_i, \quad n = \sum_{1}^{s} n_i.$$

Meinert and McHugh [1968] propose the following termination rule for the iterative processes:

$$(\hat{\theta}_{1,r} - \hat{\theta}_{1,r-1})^2 + (\hat{\theta}_{2,r}^{-1} - \hat{\theta}_{2,r-1}^{-1})^2 \leq 10^{-8}.$$

Confidence limits for x_i, the expected concentration of unlabelled antigen in the i-th set of tubes, can be computed in two steps. First, the lower and upper confidence interval limits for η_i are obtained as follows:

$(\bar{y}_{i.})_L = \bar{y}_{i.} - t_{n-2, 1-\alpha/2}(\text{var}\,\bar{y}_{i.})^{1/2}$

$(\bar{y}_{i.})_U = \bar{y}_{i.} + t_{n-2, 1-\alpha/2}(\text{var}\,\bar{y}_{i.})^{1/2}$

where $\text{var}(\bar{y}_i)$ is the estimated variance function, which is obtained numerically (an expression for it will be given later), and $\bar{y}_{i.}$ is the mean proportion of bound antigen observed in the number of test tubes containing the i-th unknown test preparation.

In the second step, the confidence limits for x_i are calculated by substituting $(\bar{y}_{i.})_L$ or $(\bar{y}_{i.})_U$ for η_i in (8) yielding:

$$(x_i)_U = \{\hat{\theta}_1\hat{\theta}_2 - (\bar{y}_{i.})_U[1 - (\bar{y}_{i.})_L]^{-1}\}/\hat{\theta}_2(\bar{y}_{i.})_L - A_L$$
$$(x_i)_L = \{\hat{\theta}_1\hat{\theta}_2 - (\bar{y}_{i.})_U[1 - (\bar{y}_{i.})_U]^{-1}\}/\hat{\theta}_2(\bar{y}_{i.})_U - A_L.$$
(9)

Example

Data generated in an insulin immunoassay [Yalow and Berson, 1960, 1970]:

Concentration of unlabelled insulin of standards, ng/ml	Observed proportion of labelled insulin bound
0	0.581
0.008	0.550
0.020	0.521
0.04	0.515
0.10	0.425
0.20	0.348
0.40	0.264
0.6	0.232
0.8	0.174
1.4	0.114

The computer routine of McHugh and Meinert [1970] gave the starting values

$\hat{\theta}_{1,0} = 0.1795$ ng/ml

$\hat{\theta}_{2,0} = 9.6297$ ml/ng.

After three iterations, the following least-square estimates were obtained.

$\hat{\theta}_1 = \hat{\theta}_{1,3} = 0.1641$ ng/ml

$\hat{\theta}_2 = \hat{\theta}_{2,3} = 11.3125$ ml/ng,

which satisfies the termination rule:

$(\hat{\theta}_{1,3} - \hat{\theta}_{1,2})^2 + (\hat{\theta}_{2,3}^{-1} - \hat{\theta}_{2,2}^{-1})^2 \leqslant 10^{-8}$.

From these estimates, the estimated standard curve can be obtained by replacing θ_1 and θ_2 by $\hat{\theta}_1$ and $\hat{\theta}_2$ in (6). Furthermore, (7) can be used to calculate $\hat{x}_i$, the unknown insulin concentration, by knowing the observed proportion of counts of bound insulin, y_i and the estimates of θ_1 and θ_2. As an example, suppose that $y_i = 0.425$ is the observed proportion of count of bound insulin with

$\hat{\theta}_1 = 0.1641$ ng/ml $y_i = 0.425$

$\hat{\theta}_2 = 11.3125$ ml/mg $A_L = 0.1$ ng/ml.

So

$$\hat{x}'_i = \frac{(0.1641)(11.3125) - [0.425/(1 - 0.425)]}{(11.3125)(0.425)} = 0.1324 \text{ ng/ml},$$

and the 95% confidence limits are

$(x'_i)_L = 0.078$ ng/ml

$(x'_i)_U = 0.194$ ng/ml.

In this case, McHugh and Meinert [1970] were able to fit a curve that was based on a knowledge of the equation for the chemical reaction that was involved. They also state that in many cases when the biochemical theory is not so well known, they use an empirical function, such as the logistic for their curve-fitting. As a comparison, these same data were fitted using a logistic regression on SAS. The mean-square error obtained for the logistic model is 0.0135327, whereas for the McHugh and Meinert model it is 0.000445. For the logistic model F is 65.77 with $P < 0.0001$ and $r^2 = 0.891558$.

11.3. Analysis of Variance of the McHugh and Meinert Model

Source	d.f.	Sum of squares	Mean square
Model	2	0.281283	1406.415×10^{-4}
Error	8	0.003567	4.45875×10^{-4}
Total	10	0.28485	284.85×10^{-4}

F ratio = MS(regression)/MS(error) = 1406.415/4.45875 = 315.4281 > $F_{0.005, 2, 8} = 11.0$.

Thus the model is highly significant at $\alpha < 0.0001$.

Remarks. The logistic regression model can also provide a good fit for the data at a highly significant level. It is reassuring to note that an empirical model which is commonly used in practice, namely the logistic, compares favorably with the model based on the theory of McHugh and Meinert, which has slightly smaller mean-square errors than the logistic model. In their model θ_1 and θ_2 have a physical interpretation, namely θ_1 representing the initial concentration of antibody and θ_2 denoting the ratio of the concentration of bound antibody-antigen complex to the product of the concentrations of free antibody and antigen at equilibrium. Also, it is worthwhile to point out that a model with a theoretical basis provides a more stable basis for the design of future experiments.

The confidence interval for x_i given by (9) depends upon the variance of $\bar{y}_i$. The approximation to this function is:

$$\widehat{\text{var}}(\bar{y}_{i\cdot}) = \left[q_i^{-1} + \sum_{k=1}^{2} \sum_{m=1}^{2} f_k(x_i';\hat{\theta}) f_m(x_i';\hat{\theta}) S^{km} \right] \hat{\sigma}^2$$

where q_i is the number of times the i-th unknown is replicated, and where

$f_k(x_i';\theta) = \partial f(x_i';\theta)/\partial \theta_k \quad f(x_i';\theta) = \eta_i = E(y_{ij})$,

$f_1(x_i';\theta) = (1/2x_i')\{1 - (\theta_1 - \theta_2^{-1} - x_i')[(\theta_1 - \theta_2^{-1} + x_i')^2 - 4\theta_1 x_i']^{-1/2}\}$

$f_2(x_i';\theta) = -(\theta_2^{-2}/2x_i')\{(\theta_1 - \theta_2^{-1} + x_i')[(\theta_1 - \theta_2^{-1} + x_i')^2 - 4\theta_1 x_i']^{-1/2} - 1\}$

$\hat{\sigma}^2 = \sum_{i=1}^{s} \sum_{j=1}^{n_i} (y_{ij} - \hat{\eta}_i)^2/(n-2), \quad n = n_1 + \cdots + n_s,$

and S^{km} is the km-th element in the inverse of the matrix (S_{km}) where

$$S_{km} = \sum_{i=1}^{s} [f_k(x_i';\hat{\theta}) f_m(x_i';\hat{\theta})] n_i \quad (k = 1, 2 \text{ and } m = 1, 2).$$

From these expressions, we infer that various ways can be found to increase the precision of the assay; they include
(1) increase q_i, the number of times the i-th unknown is replicated,
(2) increase s, the number of standards,
(3) increase n, the total number of tubes upon which the calibration curve is based.

11.4. Other Models for Radioimmunoassays

The kinetic relations of enzyme chemistry, which have a lot in common with the antigen-antibody, can often be reduced to rectangular hyperbolae. Bliss [1970] fits a three-parameter hyperbola (two asymptotes and a constant) to the measurements of human luteinizing hormone presented by Rodbard et al. [1970]. The hyperbola is given by

$$(x' - x_0')(y' - y_0') = x_0' y_0' + c \tag{10}$$

where x_0' and y_0' denote the asymptotes, c is a constant, x', the independent variable is the dose and y' is the dependent variable or the response, in its initial form. We assume that the dose is free of random error. We can convert the hyperbola into a straight line

$$Y = y_0' + \frac{x_0' y_0' + c}{x' - x_0'} = a' + b(x' + d)^{-1} = a' + bx \tag{11}$$

where $d = -x'_0$ and $x = (x' + d)^{-1}$. Since this equation is nonlinear in its parameters, its solution is indirect and stepwise.

Bliss [1970] proposes obtaining a preliminary estimate of d as follows: the values of Y are plotted against k values of $\log x'$ and fitted with a smooth curve drawn by free hand. At three equally and widely spaced levels of $\log x'_1$, $\log x'_2$ and $\log x'_3$ the corresponding values of y_1, y_2, and y_3 are interpolated from the curve. Let

$$h = (x'_3 - x'_2)/(x'_2 - x'_1). \tag{12}$$

Then, the initial value of d is computed as

$$d_0 = [hx'_1 y_1 - (h + 1)x'_2 y_2 + x'_3 y_3]/[(h + 1)y_2 - hy_1 - y_3]. \tag{13}$$

(One could average over such values of d_0 obtained by selecting different sets of equally and widely spaced levels of $\log x'_1$, $\log x'_2$ and $\log x'_3$.)

The initial estimate of the statistic d, with which the doses are transformed to $x = 100/(x' + d)$, can be improved by a simple adjustment. For each of several values of d_i, the slope of the response y upon the transformed dose x is computed by least squares and also the standard deviation $\tilde{s}$ for the scatter in y about each line. The improved value of d is that one for which $\tilde{s}$ is a minimum. When once d is chosen, we can estimate a' and b by the linear regression method. Bliss [1970] illustrates this method by the data of Rodbard et al. [1970].

Rodgers [1984] has written a survey paper on data analysis and quality control of radioimmunoassays. This will be reviewed here.

11.5. *Assay Quality Control*[1]

Recall that a plot of the response versus the analyte concentration (dose) constitutes a calibration curve which represents the quantitative relationship between what we observe, namely, the response, and what we wish to estimate, namely, the concentration of analyte. Using the calibration curve in order to determine the analyte concentration that corresponds to a specified response is called the analyte concentration interpolation, or simply interpolation. Calibration and interpolation are the core aspects of bioassay. However, sometimes, the assay concentration estimates are subject to error; this error must be quantified and controlled. This is the task of assay quality control. The error in dose levels (or analyte concentrations) can be a random or bias (systematic) error. Random error can be controlled to a limited extent by replication of a given measurement. Bias error poses a more difficult problem. Spot samples (also called quality control samples) constitute the main source of information about bias error in an assay.

---

[1] Butt [1984] served as a source for sections 11.5–11.8.

11.6. Counting Statistics

The response, for instance, in radioimmunoassays is in the form of counts of radioactive decays. As events dispersed randomly in time, these counts behave like Poisson variables. If the number of counts (including background) in the bound fraction of a given sample of a given assay is B, then

$$\text{var}(B) \doteq B, \quad \text{s.d.}(B) = B^{1/2} \quad \text{and} \quad \text{c.v.}(B) \doteq (B^{1/2}/B) = B^{-1/2}.$$

Many assayists use the following sampling rule: stop counting in each assay tube until a preset number of counts or preset counting time has been reached, whichever is realized first. Usually, the preset number of counts is 10,000. Then the standard deviation of such a count is 100. However, if one follows a more rational approach to counting, it can result in a 2- to 17-fold improvement in the efficiency of counting device utilization. Under some statistical assumptions, the error in an assay concentration estimate is proportional to the error in the corresponding response. Hence, we can confine to controlling the error in the assay response. Let the response be

$$b = B/Tt$$

where T = total counts and t = counting time used for obtaining B counts in the bound fraction. Then

$$\text{var}(b) \doteq b/tT, \quad \text{s.d.}(b) \doteq B^{1/2}/tT, \quad \text{c.v.}(b) \doteq B^{-1/2}.$$

Notice that the standard deviation of b is much smaller than that of B and the coefficients of variation of b and B are equal.

This reexpression of counting error in terms of the response, although helpful, does not directly improve our counting strategy. If s_c^2 denotes the variance in b due to counting and s_e^2 denotes the variance due to experimental error, then we can write the total variance in b as

$$s^2 = s_e^2 + s_c^2.$$

Further, often we have the relationship

$$s_e^2 = \alpha + \beta b$$

where α and β are some constants. Further, if f is such that $s^2 = (f^2 - 1)s_e^2$, then the appropriate counting time t may be defined by:

$$t = \{T(f^2 - 1)[(2/b) + \beta]\}^{-1}.$$

11.7. Calibration Curve Fitting

More attention has been given to presenting new formulae for curve fitting than to good statistical and computing procedure. Two basic problems are: (1) to find an appropriate function for the calibration curve, and (2) to fit it correctly. Although there are a large number of calibration formulae that have been proposed, they fall into three categories: (a) empirical, (b) semi-empirical and (c) model-based.

Empirical Methods

These are called empirical because their use is not based on some psysicochemical model for the assay, and they have been very successful in practice. Polynomials, splines and polygonals are examples of such empirical fits.

Semi-Empirical Methods

The techniques are called semi-empirical because they have some theoretical justification why they should fit calibration data. The most popular method is the logistic technique. The four-parameter logistic response curve, introduced by Heally [1972], is given by

$$y = (a - d)[1 + (x/c)^b]^{-1} + d,$$

a, b, c and d being the unknown parameters. Parameters a and d correspond to the upper and lower asymptotes of the curve, respectively: c is the value of x (analyte concentration) which corresponds to the center or inflection point, and b is related to the slope of the center of the curve. Also, one can rewrite the curve as

$$\ln[(y - d)/(a - y)] = (-b)\ln x + b \ln c$$

The logit transformation is

$$Y = \ln[(y - d)/(a - y)].$$

So, one can fit a straight line Y versus $\ln x$ with a slope of $-b$ and Y intercept of $b \ln c$.

Rodbard et al. [1969] were the first to fit a logistic function to binder-log and assay (this method is known as the logit-log technique) by setting $a = 1$ and $d = 0$. The value used for y is the bound counts for a given analyte concentration divided by the bound counts at zero analyte concentration (after correcting all bound count values for the effects of nonspecific binding).

The practical difficulty with the four-parameter logistic method is that it is nonlinear in its parameters b and c. However, Zivitz and Hidalgo [1977] proposed a strategy by which linear regression may still be applied to the four parameter logistic [Davis et al., 1980]. If one has good estimates of a and d based on previous experience, the logit transformation can be performed on the calibration curve data and a straight line fitted to the transformed data, using weighted linear least squares. This yields estimates for b and c. One can then calculate the value of $x' = [1 + (x/c)^b]^{-1}$ for each datum and fit a straight line to y versus x'. This yields new estimates for a and d. Iterate this process until the two linear curve-fitting procedures converge to a desirable accuracy. The most difficult component of this simple procedure is the suitable determination of weights to be used for the two linear fitting procedures. The above method is known as 2 + 2 logistic technique.

Notice that the four-parameter logistic model is symmetrical about a central inflection point. If this is not appropriate for the data, then a generalized logistic can be used. Note that the four-parameter logistic is of the form

$$y = (a - d)\{1 + \exp[f(\ln x)]\}^{-1} + d, \quad \text{where}$$

$$f(\ln x) = b \ln x - \ln c.$$

Now, a generalized logistic is obtained by setting f to be an arbitrary function. If f is a polynomial, a 2 + n logistic method could be followed in analogy to the 2 + 2 technique, using only linear regression methods. Here n denotes the order of the polynomial + 1, or the number of parameters to be fitted to describe the polynomial chosen. Usually $n \leqslant 4$ is adequate. One should make sure that the resulting curve is monotonic. Also, one can employ one of the nonlinear curve-fitting computer programs which are readily accessible.

Model-Based Methods

The final group of calibration curve-fitting methods is based on simple chemical models for the assay reactions in use. A sample scheme for the central reactions of binder-ligand radioimmunoassay is

$$P + Q \underset{k_r}{\overset{k_f}{\rightleftharpoons}} PQ$$

$$P^* + Q \underset{k_r^*}{\overset{k_f^*}{\rightleftharpoons}} PQ$$

where P^* denotes the labeled ligand, P the unlabelled ligand (the analyte) and Q denotes the binder, often the antibody. These two competing reactions are assumed to be governed by the simple first-order mass action law:

$$K = \frac{k_f}{k_r} = \frac{[PQ]}{[P][Q]} \quad \text{and} \quad K^* = \frac{k_f^*}{k_r^*} = \frac{[P^*Q]}{[P^*][Q]}$$

where the square brackets indicate the concentrations of the contained species at chemical equilibrium, and the K values are the equilibrium constants describing these two reactions. Let q denote the total concentration of binder, and p^* and p are the total concentrations of labeled and unlabeled ligands, respectively. An example of such a model is that of Naus et al. [1977]. Some presently available models assume that $K = K^*$. Others add a parameter B_n to describe nonspecific binding, which is often assumed to be a set fraction of the ligand present (an assumption which is rarely validated). Several models allow for a multiplicity of binding sites (binder heterogeneity). Raab [1983] and Finney [1983] have demonstrated the practical defects in model-based calibration curve-fitting.

11.8. *Principles of Curve-Fitting*

In the early years, manual techniques were used in immunoassays. Calibration results were plotted and a curve was drawn by hand through these points. However, due to compelling reasons, a statistical fit is preferable. For automated curve-fitting, the most commonly used technique is the method of least squares which makes the following assumptions: (1) the x values are free of error; (2) the formula selected for fitting truly describes the data, and (3) that there are no outliers in the data. Besides, it would be helpful if the errors in the y values are normally distributed. The method of least squares could be linear or nonlinear. The number of parameters in the model plays an important role in determining how flexible a given equation will be. For example, a four-parameter polynomial will be able to fit a wider variety of shapes than a straight line. It is often said jokingly that a four-parameter model is used to describe an elephant, and with a fifth parameter the elephant can be made to wag its tail. By Anova techniques, one can test the validity of the model.

The problems one is faced with are: (1) the random errors in the values tend to vary as a function of y (and consequently as a function of x), and (2) the outliers. The problem of heteroscedasticity can be overcome by using weighted least squares regression methods. However, the rejection of outliers has never been satisfactorily resolved. Robust regression methods are desirable in this regard. For instance, the method of Tiede and Pagano [1979] has been criticized on several counts by Raab [1981]. The assayist's judgement should be brought into play regarding outliers. The presence of a large number of outliers is a sign of a loss of quality control.

Rodgers [1984] discusses how to overcome other errors in assays, such as bias (or systematic) error, and the assay design. Rodgers [1984] recommends using automated packages rather than writing one's own program. We should also be wary of commercial software. Select a program that represents the current state of the art, such as the Edinburgh FORTRAN program [McKenzie and Thompson, 1982], the NIH FORTRAN program [Rodbard et al., 1975]. Rodgers [1984] recommends establishing a comprehensive internal quality control program.

References

Aitchison, J.; Silvey, S. D.: The generalization of probit analysis to the case of multiple responses. Biometrika 44: 131–140 (1957).
Amemiya, J.: The n^{-2} order mean squared errors of the maximum likelihood and the minimum logit chi-square estimator. Ann. Statist. 8: 488–505 (1980).
Anscombe, F. J.: On estimating binomial response relations. Biometrika 43: 461–464 (1956).
Antoniak, C. E.: Mixtures of Dirichlet processes with applications to Bayesian nonparametric problems. Ann. Statist. 2: 1153–1174 (1974).
Armitage, P.; Doll, R.: Stochastic models for carcinogenisis. Proc. 4th Berkeley Symp. Math. Statist. Prob., vol. 4, pp. 19–38 (University of California Press, Berkeley 1961).
Ashford, J. R.: An approach to the analysis of data for semiquantal responses in biological assay. Biometrics 15: 573–581 (1959).
Ashton, W.: The logit transformation. With special reference to its uses in bioassay (Hafner, New York 1972).
Aspin, A. A.: An examination and further development of a formula arising in the problem of comparing two mean values. Biometrika 35: 88–96 (1948).
Aspin, A. A.: Tables for use in comparisons whose accuracy involves two variances. Biometrika 36: 290–296 (1949).
Ayer, M.; Brunk, H. D.; Ewing, G. M.; Reid, W. T.; Silverman, E.: An empirical distribution function for sampling with incomplete information. Ann. math. Statist. 26: 641–647 (1955).
Barlow, R. E.; Bartholomew, D. J.; Bremer, J. M.; Brunk, H. D.: Statistical inference under order restrictions (Wiley, New York 1972).
Barlow, W. E.; Feigl, P.: Fitting probit and logit models with nonzero background using GLIM (unpublished manuscript, 1982).
Barlow, W. E.; Feigl, P.: Analyzing binomial data with a non-zero baseline using GLIM. Comput. Statist. Data Analy. 3: 201–204 (1985).
Barnard, G.: The Behrens-Fisher test. Biometrika 37: 203–207 (1950).
Bartlett, M. S.: Properties of sufficiency and statistical tests. Proc. R. Soc. A160: 268–282 (1937).
Bartlett, M. S.: The use of transformations. Biometrics 3: 39–52 (1947).
Behrens, W. U.: Ein Beitrag zur Fehler-Berechnung bei wenigen Beobachtungen. Landw Jb. 68: 807–837 (1929).
Berkson, J.: A statistically precise and relatively simple method of estimating the bioassay with quantal response, based on the logistic function J. Am. Statist. Assoc. 48: 565–599 (1953).
Berkson, J.: Maximum likelihood and minimum chi-square estimates of the logistic function. J. Am. Statist. Ass. 50: 130–162 (1955).
Berkson, J.: Tables for use in estimating the normal distribution function by normit analysis. Biometrika 44: 441–435 (1957a).
Berkson, J.: Tables for the maximum likelihood estimate of the logistic function. Biometrics 13: 28–34 (1957b).
Berkson, J.: Smoking and lung cancer: some observations on two recent reports. J. Am. Statist. Ass. 53: 28–38 (1958).
Berkson, J.: Application of minimum logit χ^2 estimate to a problem of Grizzle with a notation on the problem of no interaction. Biometrics 24: 75–95 (1968).
Berkson, J.: Minimum chi-square, not maximum likelihood! Ann. Statist. 8: 447–487 (1980).

References

Biggers, J. D.: Observations on the intravaginal assay of natural oestrogens using aqueous egg albumin as the vehicle of administration. J. Endocr. 7: 163–171 (1950).

Bishop, Y. M. M.; Fienberg, S. E.; Holland, P. W.: Discrete multivariate analysis: theory and practice, pp. 401–433 (MIT Press, Cambridge 1975).

Bliss, C. I.: The method of probits. Science 79: 409–410 (1934).

Bliss, C. I.: The calculation of the dosage-mortality curve. Ann. app. Biol. 22: 134–167 (1935).

Bliss, C. I.: The analysis of field experimental data expressed in percentages. Plant Protect. 12: 67–77 (1937).

Bliss, C. J.: The statistics of bioassay (Academic Press, New York 1952).

Bliss, C. I.: Dose-response curves for radioimmunoassays; in McArthur, Colton, Statistics in endocrinology, pp. 399–410 (MIT Press, Cambridge 1970).

Block, H. D.: Estimates of error for two modifications of the Robbins-Munro stochastic approximation process. Ann. math. Statist. 28: 1003–1010 (1957).

Blum, J. R.: Multidimensional stochastic approximation methods. Ann. math. Statist. 25: 737–744 (1954).

Bradley, R. A.; Gart, J. J.: The asymptotic properties of ML estimators when sampling from associated populations. Biometrika 49: 205–214 (1962).

Brand, R. J.; Pinnock, D. E.; Jackson, K. L.: Large sample confidence bands for the logistic response curve and its inverse. Am. Statist. 27: 157–160 (1973).

Brown, B. W., Jr.: Some properties of the Spearman estimator in bioassay. Biometrika 48: 293–302 (1961).

Brown, B. W., Jr.: Planning a quantal assay of potency. Biometrics 22: 322–329 (1966).

Brown, B. W., Jr.: Quantal response assays; in McArthur, Colton, Statistics in endocrinology, pp. 129–143 (MIT Press, Cambridge 1970).

Brownless, K. A.; Hodges, J. L.; Rosenblatt, M.: The up and down method with small samples. J. Am. Statist. Ass. 48: 262–277 (1953).

Burn, J. H.; Finney, D. J.; Goodwin, L. G.: Biological standardization (Oxford University Press, London 1950).

Butt, W. R. (ed.): Practical immunoassay: the state of the art, pp. 253–368 (Marcel Dekker, New York 1984).

Cheng, P. C.; Johnson, E. A.: Some distribution-free properties of the asymptotic variance of the Spearman estimator in bioassays. Biometrics 28: 882–889 (1972).

Chernoff, H.: Asymptotic studentization in testing hypotheses. Ann. math. Statist. 20: 268–278 (1949).

Chmiel, J. J.: Some properties of Spearman-type estimators of the variance and percentiles in bioassay. Biometrika 63: 621–626 (1976).

Choi, S. C.: An investigation of Wetherill's method of estimation for the up and down experiment. Biometrics 27: 961–970 (1971).

Chung, K. L.: On a stochastic approximation method. Ann. math. Statist. 25: 463–483 (1954).

Church, J. D.; Cobb, E. B.: On the equivalence of Spearman-Karber and maximum likelihood estimates of the mean. J. Am. Statist. Ass. 68: 201–202 (1973).

Clark, A. J.: The mode of action of drugs on cells, p. 4 (Edward Arnold, London 1933).

Cochran, W. G.; Davis, M.: Stochastic approximation to the median effective dose in bioassay. Stochastic model in medicine and biology, pp. 281–300 (University of Wisconsin Press, Madison 1964).

Colguhown, D.: Lectures in biostatistics. An introduction to statistics with applications in biology and medicine (Clarendon Press, Oxford 1971).

Copenhaver, T. W.; Mielke, P. W.: Quantit analysis: a quantal assay refinement. Biometrics 33: 175–186 (1977).

Cornfield, J.: A statistical problem arising from retrospective studies. Proc. 3rd Berkeley Symp. Math. Statist. Prob., vol. 4, pp. 135–148 (University of California Press, Berkeley 1956).

Cornfield, J.; Mantel, N.: Some new aspects of the application of maximum likelihood to the calculation of the dosage response curve. J. Am. Statist. Ass. 45: 181–201 (1956).

Coward, K. H.: The biological standardization of the vitamin; 2nd ed. (Baillière, Tindall & Cox, London 1947).
Cox, D. R.: Some procedures connected with the logistic quantitative response curve; in David, Research papers in statistics, Festschrift for J. Neyman, pp. 55–71 (Wiley, New York 1966).
Cox, D. R.: The analysis of binary data (Methuen, London 1970).
Cox, D. R.: Regression models and life tables. J. R. Statist. Soc. Ser. B *34:* 187–202 (1972).
Cramer, E. M.: Some comparisons of methods of fitting the dosage response curve for small samples. J. Am. statist. Ass. *59:* 779–793 (1964).
Creasy, M. A.: Limits for the ratio of means. J. R. Statist. Soc. Ser. B *16:* 186–192 (1954).
Crump, K. S.; Guess, H. A.; Deal, K. L.: Confidence intervals and tests of hypotheses concerning dose response relations inferred from animal carcinogenicity data. Biometrics *33:* 437–451 (1977).
Davis H. T.: The analysis of economic time series (Trinity University Press, San Antonio 1941).
Davis, S. E.; Jaffe, M. L.; Munson, P. J.; Rodbard, D.: RIA data processing with a small programmable calculator. J. Immunoassay *1:* 15–25 (1980).
Derman, G.: Stochastic approximation. Ann. math. Statist. *27:* 879–886 (1956).
Dixon, W. J.: The up and down method for small samples. J. Am. Statist. Ass. *60:* 967–978 (1965).
Dixon, W. J.: Quantal response variable experimentation: the up and down method; in McArthur, Colton, Statistics in endocrinology, pp. 251–267 (MIT Press, Cambridge 1970).
Dixon, W. J.; Mood, A. M.: A method for obtaining and analysing sensitivity data. J. Am. Statist. Ass. *43:* 109–126 (1948).
Dorn, H. F.: The relationship of cancer of the lung and the use of tobacco. Am. Statist. *8:* 7–13 (1954).
Draper, N. R.; Smith, H.: Applied regression analysis; 2nd ed., sect. 6.3, 6.4 (Wiley, New York, 1981).
Dvoretzky, A.: On stochastic approximation. Proc. 3rd Berkeley Symp. Math. Statist. Prob., vol. 1, pp. 39–55 (University of California Press, Berkeley 1956).
Emmens, C. W.: The dose/response relation for certain principles of the pituitary gland, and of the serum and urine of pregnancy. J. Endocr. *2:* 194–225 (1940).
Emmens, C. W.: Principles of biological assay (Chapman & Hall, London 1948).
Epstein, B.; Churchman, C. W.: On the statistics of sensitivity data. Ann. math. Statist. *15:* 90–96 (1944).
Feinstein, A. R.: Clinical biostatistics. XX. The epidemiologic trohoc, the ablative risk ratio, and retrospective research. Clin. Pharmacol. Ther. *14:* 291–307 (1973).
Ferguson. T. S.: A Bayesian analysis of some nonparametric problems. Ann. Statist. *1:* 200–230 (1973).
Fieller, E. C.: The biological standardization of insulin. J. R. Statist. Soc., suppl. 7, pp. 1–64 (1940).
Fieller, E. C.: A fundamental formula in the statistics of biological assay, and some applications. Q. J. Pharm. Pharmacol. *17:* 117–123 (1944).
Fieller, E. C.: Some problems in interval estimation. J. R. Statist. Soc. Ser. B *16:* 175–186 (1954).
Finney, D. J.: The principles of biological assay. J. R. Statist. Soc., suppl. 9, pp. 46–91 (1947).
Finney, D. J. The estimation of the parameters of tolerance distributions. Biometrika *36:* 239–256 (1949).
Finney, D. J.: The estimation of the mean of a normal tolerance distribution. Sankhya *10:* 341–360 (1950).
Finney, D. J.: Two new uses of the Behrens-Fisher distribution. J. R. Statist. Soc. Ser. B *12:* 296–300 (1951).
Finney, D. J.: Probit analysis, a statistical treatment of the sigmoid response curve; 2nd ed. (Cambridge University Press, London 1952).
Finney, D. J.: The estimation of the ED_{50} for logistic response curve. Sankhya *12:* 121–136 (1953).
Finney, D. J.: Probit analysis; 3rd ed. (Cambridge University Press, London 1971a).
Finney, D. J.: Statistical method in biological assay; 2nd ed. (Griffin, London 1971b).
Finney, D. J.: Response curves for radioimmunoassay. Clin. Chem. *29:* 1562–1566 (1983).
Fisher, R. A.: The case of zero survivors. Ann. appl. Biol. *22:* 164–165 (1935a).
Fisher, R. A.: The fiducial argument in statistical inference. Ann. Eugen. *6:* 391–398 (1935b).

Fisher, R. A.: The asymptotic approach to Behrens integral with further tables for the d test of significance. Ann. Eugen. *11:* 141–172 (1941).
Fisher, R. A.: Sampling the reference set. Sankhya Ser. A *23:* 3–8 (1961a).
Fisher, R. A.: The weighted mean of two normal samples with unknown variance ratio. Sankhya Ser. A *23:* 103–144 (1961b).
Fisher, R. A.; Yates, F.: Statistical tables for biological, agricultural and medical research; 6th ed. (Oliver & Boyd, Edinburgh 1963).
Fleiss, J.: Statistical methods for rates and proportions; 2nd ed. (Wiley, New York 1981).
Freeman, P. R.: Optimal Bayesian sequential estimation of the median effective dose. Biometrika *57:* 79–89 (1970).
Freundlich, H.: Colloid and capillary chemistry, p. 141 (Dutton, New York 1922).
Gaddum, J. H.: Pharmacology; 3rd ed. (Oxford University Press, London 1948).
Galton, F.: The geometric mean in vital and social statistics. Proc. R. Soc. *29:* 365–367 (1879).
Gart, J. J.: Point and interval estimation of the common odds ratio in the combination of 2×2 tables with fixed marginals. Biometrika *57:* 471–475 (1970).
Gart. J. J.; Zweifel, J. R.: On the bias of various estimators of the logit and its variance with applications to quantal bioassay. Biometrika *54:* 181–187 (1967).
Garwood, F.: The application of maximum likelihood to dosage-mortality curves. Biometrika *32:* 46–58 (1941).
Grizzle, J. E.: A new method of testing hypotheses and estimating parameters for the logistic model. Biometrics *17:* 372–385 (1961).
Govindarajulu, Z.; Lindqvist, B. H.: Asymptotic efficiency of the Spearman estimator and characterizations of distributions. Ann. Inst. Statist. Math., Tokyo *39A:* 349–361 (1987).
Guess, H. A.; Crump, K. S.: Low dose-rate extrapolation of data from animal carcinogenicity experiments: analysis of a new statistical technique. Math. Biosci. *30:* 15–36 (1976).
Guess. H.; Crump, K.: Can we use animal data to estimate safe doses for chemical carcinogens? in Whittmore, Environmental health: quantitative methods, pp. 13–28 (Society for Industrial and Applied Mathematics, Philadelphia 1977).
Guess, H. A.; Crump, K. S.: Maximum likelihood estimation of dose-response functions subject to absolutely monotonic contraints. Ann. Statist. *6:* 101–111 (1978).
Guess, H.; Crump, K.; Peto, R.: Uncertainty estimates for low dose rate extrapolation of animal carcinogenecity data. Cancer Res. *37:* 3475–3483 (1977).
Gurland, J.; Lee, I., Dahm, P.: Polychotomous quantal response in biological assay. Biometrics *16:* 382–398 (1960).
Hartley, H. O. The modified Gauss-Newton method for the fitting of nonlinear regression functions by least squares. Technometrics *3:* 269–280 (1961).
Hartley, H. O.; Sielkin, R. L.: Estimation of 'safe dose' in carcinogenic experiments. Biometrics *33:* 1–30 (1977).
Hauck, W. W.: A note on confidence bands for the logistic response curve. Am. Statist. *37:* 158–160 (1983).
Heally, M. J. R.: Statistical analysis of radioimmunoassay data. Biochem. J. *130:* 207–210 (1972).
Hodges, J. L. Jr.: Fitting the logistic by maximum likelihood. Biometrics *14:* 453–461 (1958).
Hodges, J. L., Jr.; Lehmann, E. L.: Two approximations to the Robbins-Monro process. Proc. 3rd Berkeley Symp. Math. Statist. Prob., vol. 1, 95–104 (University of California Press, Berkeley 1956).
Hoel, P. G.; Jennrich, R. I.: Optimal designs for dose-response experiments in cancer research. Biometrika *66:* 307–316 (1979).
Hoel, P. G.: A simple solution for optimal Chebyshev regression extrapolation. Ann. math. Statist. *37:* 720–725 (1966).
Hsi, B. P.: The multiple sample up and down method in bioassay. J. Am. Statist. Ass. *64:* 147–162 (1969).
Hsu, P. L.: Contributions to the theory of Student's t-test as applied to the problem of two samples. Statist. Res. Mem. *2:* 1–24 (1938).

James, A. T.; Wilkinson, G. N.; Venables, W. W.: Interval estimates for a ratio of means. Sankhya Ser. A. *36:* 177–183 (1974).
Jerne, N. K.; Wood, E. C.: The validity and meaning of the results of biological assays. Biometrics *5:* 273–299 (1949).
Johnson, E. A.; Brown, B. W., Jr.: The Spearman estimator for serial dilution assays. Biometrics *17:* 79–88 (1961).
Kraft, C. H.; Eeden, C. van: Bayesian bioassay. Ann. math. Statist. *35:* 886–890 (1964).
Kundson, L. F.; Curtis, J. M.: The use of the angular transformation in biological assays. J. Am. Statist. Ass. *42:* 282–296 (1947).
Langmuir, I.: The shapes of group molecules forming the surfaces of molecules. Proc. natn. Acad. Sci. USA *3:* 251–257 (1917).
LeCam, L.: An extension of Wald's theory of statistical decision functions. Ann. math. Statist. *26:* 69–81 (1955).
Lindgren, B. W.: Statistical theory; 3rd ed. (McMillan, New York 1976).
Little, R. E.: A note on estimation for quantal response data. Biometrika *55:* 578–579 (1968).
Maggio, E. G.: Enzyme-immunoassay (CRC Press, Boca Raton 1980).
Magnus, A.; Mielke, P. W.; Copenhaver, T. W.: Closed expression for the sum of an infinite series with application to quantal response assays. Biometrics *33:* 221–224 (1977).
Mantel, N.; Haenszel, W.: Statistical aspects of the analysis of data from retrospective studies of disease. J. natn. Cancer Inst. *22:* 719–748 (1959).
Mantel, N.; Bryan, W. R.: 'Safety' testing of carcinogenic agents. J. natn. Cancer Inst. *27:* 455–470 (1961).
Mantel, N.; Bohidar, N.; Brown, C.; Ciminera, J.; Tukey, J.: An improved Mantel-Bryan procedure for 'safety' testing of carcinogens. Cancer Res. *35:* 865–872 (1975).
Marks, B. L.: Some optimal sequential schemes for estimating the mean of a cumulative normal quantal response curve. J. R. Statist. Soc. Ser. B *24:* 393–400 (1962).
McHugh, R. B.; Meinert, C. L.: A theoretical model for statistical inference in isotope displacement immunoassay; in McArthur, Colton, Statistics in endocrinology, pp. 399–410 (MIT Press, Cambridge 1970).
McKenzie, G. M.; Thompson, R. C. H.: Design and implementation of a software package for analysis of immunoassay data; in Hunter, Corrie, Immunoassays for clinical chemistry. A Workshop Meeting (Churchill Livingstone, Edinburgh 1982).
McLeish, D. L.; Tosh, D. H.: The estimation of extreme quantiles in logit bioassay. Biometrika *70:* 625–632 (1983).
McLeish, D. L.; Tosh, D. H.: Two-dose allocation schemes in logit analysis with cost restraints (unpublished manuscript, 1985).
Meinert, C. L.; McHugh, R. B.: The biometry of an isotope displacement immunologic microassay. Math. Biosci. *2:* 319–338 (1968).
Miller, R. G.: Nonparametric estimators of the mean tolerance in bioassay. Biometrika *60:* 535–542 (1973).
Morton, J. T.: The problem of the evaluation of retenone-containing plants. VI. The toxicity of l-elliptone and of poisons applied jointly, with further observations on the retenone equivalent method of assessing the toxicity of derris root. Ann. appl. Biol. *29:* 69–81 (1942).
Naus, A. J.; Kuffens, P. S.; Borst, A.: Calculation of radioimmunoassay standard curves. Clin. Chem. *23:* 1624–1627 (1977).
Neyman, J.: Contributions to the theory of the chi-square test. Proc. Berkeley Symp. Math. Statist. Prob., pp. 239–272 (University of California Press, Berkeley 1949).
Oliver, F. R.: Methods of estimating the growth function. J. R. Statist. Soc. Ser. C *13:* 57–66 (1964).
Pearl, R.: Studies in human biology (Willams & Wilkins, Baltimore 1924).
Peto, R.; Lee, P.: Weibull distributions for continuous carcinogensis experiments. Biometrics *29:* 457–470 (1973).
Prentice, R. L.: A generalization of the probit and logit methods for dose-response curves. Biometrics *32:* 761–768 (1976).

Raab, G. M.: Robust calibration and radioimmunoassay (letter). Biometrics *37:* 839–841 (1981).
Raab, G. M.: Comparison of a logistic and a mass-action curve for radioimmunoassay data. Clin. Chem. *29:* 1757–1761 (1983).
Ramsey, F. L.: A Bayesian approach to bioassay. Biometrics *28:* 841–858 (1972).
Rao, C. R.: Linear statistical inference and its applications (Wiley, New York 1965).
Rao, C. R.: Estimation of relative potency from multiple response data. Biometrics *10:* 208–220 (1954).
Rechnitzer, P. A.; Sango, S.; Cunningham, D. A.; Andrew, G., Buck, C.; Jones, N. L.; Kavanagh, T.; Parker, J. O.; Shepherd, R. J.; Yuhasz, M. S.: A controlled prospective study of the effect of endurance training on the recurrence of myocardial infarction – a description of the experimental design. Am. J. Epidem. *102:* 358–365 (1975).
Rechnitzer, P. A.; Cunningham, D. A.; Andrew, G. M.; Buck, C. W.; Jones, N. L.; Kavanagh, T. B.; Oldridge, N. B.; Parker, J. O.; Shephard, R. J.; Sutton, J. R.; Donner, A.: The relationship of exercise to the recurrence rate of myocardial infarction in men – Ontario Exercise Heart Collaborative Study. Am. J. Cardiol. *51:* 65–69 (1983).
Richards, F. J.: A flexible growth function for empirical use. J. exp. Bot. *10:* 290–300 (1959).
Rizzardi, F. R.: (1985). Some asymptotic properties of Robbins-Monro type estimators with applications to estimating medians from quantal response; diss. University of California, Berkeley (1985).
Robbins, H.; Monro, S.: A stochastic approximation method. Ann. math. Statist. *22:* 400–407 (1951).
Rodbard, D.; Bridson, W.; Rayford, P. L.: Rapid calculation of radioimmunoassay results. J. Lab. clin. Med. *74:* 770–781 (1969).
Rodbard, D.; Faden, V. B.; Knisley, S.; Hutt, D. M.: Radio-immunoassay data processing; logit-log, logistic method, and quality control; 3rd ed., reports No. PB246222, PB246223, and PB246224 (National Technical Information Science, Springfield 1975).
Rodbard, D.; Rayford, P. L.; Ross, G. T.: Statistical quality control of radioimmunoassays; in McArthur, Colton, Statistics in endorcrinology, pp. 411–429 (MIT Press, Cambridge 1970).
Rodgers, R. P. C.: Data analysis and quality control of assays: a practical primer; in Butt, Practical immunoassay, the state of the art, pp. 253–368 (Marcel Dekker, New York 1984).
Sacks, J.: Asymptotic distribution of stochastic approximation procedures. Ann. math. Statist. *29:* 373–405 (1958).
Scheffé, H.: On solutions to the Behrens-Fisher problem based on the t-distribution. Ann. math. Statist. *14:* 35–44 (1943).
Schultz, H.: The standard error of a forecast from a curve. J. Am. Statist. Ass. *25:* 139–185 (1930).
Shepard, H. H.: Relative toxicity at high percentages of insect mortality. Nature *134:* 323–324 (1934).
Sheps, M. C.: Shall we count the living or the dead? New Engl. J. Med. *259:* 1210–1214 (1958).
Sheps, M. C.: Marriage and mortality. Am. J. publ. Hlth *51:* 547–555 (1961).
Smith, K. C.; Savin, N. E.; and Robertson, J. L.: A Monte Carlo comparison of maximum likelihood and minimum chi-square sampling distributions in logit analysis. Biometrics *40:* 471–482 (1984).
Sokal, R. R.; Rohlf, F. J.: Biometry; 2nd ed., p. 733 (Freeman, New York 1981).
Sokal, R. R.: A comparison of fitness characters and their response to density in stock and selected cultures of wild type and black *Tribolium castaneum.* Tribolium Inf. Bull. *10:* 142–147 (1967).
Solomon, L.: Statistical estimation. J. Inst. Act. Stud. Soc. *7:* 144–173 (1948a).
Solomon, L.: Statistical estimation. J. Inst. Act. Stud. Soc. *7:* 213–234 (1948b).
Spearman, C.: The method of 'right and wrong cases' ('constant stimuli') without Gauss' formulae. Br. J. Psychol. *2:* 227–242 (1908).
Spurr, W. A.; Arnold, D. R.: A short-cut method of fitting a logistic curve. J. Am. Statist. Ass. *43:* 127–134 (1948).
Stevens, W. L.: Mean and variance of an entry in a contingency table. Biometrika *38:* 468–470 (1951).
Stone, M.: Cross-validation and multinomial prediction. Biometrika *61:* 509–515 (1974).
Suits, C. G.; Way, E. E.: The collected works of Irwing Langmuir, vol. 9, Surface phenomena, pp. 88, 95, 445 (Pergamon, New York 1961).

Sukhatme, P. V.: On Fisher and Behrens test of significance for the difference in means of two normal populations. Sankhya *4:* 39–48 (1938).

Taylor, W. F.: Distance functions and regular best asymptotically normal estimates. Ann. math. Statist. *24:* 85–92 (1953).

Thomas, D. G.; Gart, J. J.: A table of exact confidence limits for differences and ratios of two proportions and their odds ratios. J. Am. Statist. Ass. *72:* 73–76 (1977).

Tiede, J. J.; Pagano, M.: The application of robust calibration to radioimmunoassay. Biometrics *35:* 567–574 (1979).

Tomatis, L.; Turusov, V.; Day, N.; Charles, R. T.: The effect of long-term exposure to DOT on CF-1 mice. Int. J. Cancer *10:* 489–506 (1972).

Tsutakawa, R. K.: Random walk design in bioassay. J. Am. Statist. Ass. *62:* 842–856 (1967).

Tsutakawa, R. K.: Design of experiments for bioassay. J. Am. statist. Ass. *67:* 584–590 (1972).

Tsutakawa, R. K.: Selection of dose levels for estimating a percentage point of a logistic quantal response curve. Appl. Satist. *29:* 25–33 (1980).

Turnbull, B. W.: Nonparametric estimation of a survivorship function with doubly censored data. J. Am. Statist. Ass. *69:* 169–173 (1974).

Turnbull, B. W.: The empirical distribution function with arbitrarily grouped, censored, and truncated data. J. R. Statist. Soc. Ser. B *38:* 290–295 (1976).

Venter, J. H.: An extension of the Robbins-Monro procedure. Ann. math. Statist. *38:* 181–190 (1967).

Wald, A.: Testing the difference between the means of two normal populations with unknown standard deviations; in Wald, Selected papers in statistics and probability, pp. 669–675 (University Press, Stanford 1955).

Walker, A. I. T.; Thorpe, E.; Stevenson, D. E.: The toxicology of dieldrin (HEOD). I. Long-term oral toxicity studies in mice. Food Cosmetics Toxicol. *11:* 415–432 (1972).

Wallace, D.: Asymptotic approximations to distributions. Ann. math. Statist. *29:* 635–654 (1958).

Welch, B. L.: The generalization of Student's problem when several different population variances are involved. Biometrika *34:* 23–35 (1947).

Wesley, M. N.: Bioassay: estimating the mean of the tolerance distribution. Stanford University Technical Report No. 17 (1 R01 GM 21215-01) (1976).

Wetherill, G. B.: Sequential estimation of quantal response curve. J. R. Statist. Soc. Ser. B *25:* 1–48 (1963).

Wetherill, G. B.: Sequential methods in statistics (Methuen, London 1966).

Wetherill, G. B.; Chen, H.; Vasudeva, R. B.: Sequential estimation of quantal response curves: a new method of estimation. Biometrika *53:* 439–454 (1966).

Wilks, S. S.: Mathematical statistics, p. 370 (Wiley, New York 1962).

Wilson, E. B.; Worcester, J.: The determination of LD_{50} and its sampling error in bioassay. Proc. natn. Acad. Sci. USA, part I, pp. 19–85, part II, pp. 114–120, part III, pp. 257–262 (1943a).

Wilson, E. B.; Worcester, J.: Bioassay on a general curve. Proc. natn. Acad. Sci. USA *29:* 150–154 (1943b).

Wolfowitz, J.: On the stochastic approximation method of Robbins and Monro. Ann. math. Statist. *23:* 457–461 (1952).

Worcester, J.; Wilson, E. B.: A table determining LD_{50} or the 50 percent end point. Proc. natn. Acad. Sci., Washington, *29:* 207–212 (1943).

Yallow, R. W.; Berson, S. A.: Plasma insulin in man (Editorial). Am. J. Med. *29:* 1–8 (1960).

Yallow, R. W.; Berson, S. A.: (1970). Radioimmunoassays; in McArthur, Colton, Statistics in endocrinology, pp. 327–344 (MIT Press, Cambridge 1970).

Yates, F.: An apparent inconsistency arising from tests of significance based on fiducial distributions of unknown parameters. Proc. Camb. Phil. Soc. *35:* 579–591 (1939).

Zivitz, M.; Hidalgo, J. V.: A linearization of the parameters in the logistic function; curve fitting radioimmunoassays. Comput. Progr. Biomed. *7:* 318 (1977).

Subject Index

Abbott's formula 40
Acetylamine fluorene 123
Adsorption 28, 29
Algorithm 119–121, 127, 138
  self-consistency 138
Antibodies 145, 146, 150
Antigens 145, 146, 148, 150
Antiserum 90–92
Analysis
  least squares 90
  logit, *see* Logit
  probit 85
  statistical 103
Analyte concentration 151, 153
Anova 19, 27, 155
Assay validity 13
Assays 1, 55, 58, 151, 153, 155
  analytical 8
  comparative 8, 13
  dilution 8, 13
  direct 1, 12,
    see also Biological assays
  estrone 80
  indirect 1, 12,
    see also Biological assays
  natriuretic 9, 10
  qualitative 1
  quantal 58
  quantitative 12
  two-point 71
Asymptotes 150
Asymptotic approximations 99
Asymptotic distribution or normality 5, 48, 53, 78, 96, 98, 100, 132
Asymptotic efficiency 47–50, 54, 102
Asymptotic optimality 128
Asymptotive variance 5, 47–49, 54, 61, 100–102, 114, 130
Asymptotically unbiased 63
Atrial natriuretic factor 1, 9, 10

Average sample number 109
  properties 112

Bacterial density 101
Bayes
  binomial estimator 139, 142
  pseudo-estimator 140
Bayesian methods 59, 134, 143
Beta prior 144
Benzopyrene 131
Behrens' distribution, *see*
  Distribution
Bias 46, 53, 59, 84, 99, 109, 110, 112, 132, 140, 145, 151
Binder heterogeneity 155
Bioassay 1, 2, 48, 72, 81, 84, 95, 138
  Bayesian 133
  quantal 133
Biochemical theory 149
Biological assays 1, 57
  types 3
  variability 145
Blum's conditions 97

Carcinogenic models 113, 117
Cat method 1
Chebyshev
  regression model 128
  system 127, 128
Coefficient of variation 54, 152
Collinearity 37
Completeness 133
Computing procedure 152
Concentration 13
Confidence bounds 77–79, 88, 105
Confidence limits or intervals
  6–10, 53, 54, 70, 80, 91, 96, 121, 126, 127, 132, 145, 148, 150
  asymptotic 126
Conjugate prior 139, 143
Consistency 48, 58

Consistent family 135
Constraint 136
Contingency table 88, 91
Convergence 94
  mean 97
  probability 94, 95
  quadratic mean 94, 95
  rate 94, 147
  strong 95, 98
Convex programming 119–121
Covariance 6, 8
Covariates 10
Criterion function 114
Cross validation 140, 141
Curve
  action 28
  calibration 147, 151–155
  dose-response 30
  growth 81
  logistic 33, 35, 44
  probit response 143
  response 44
  sigmoid 32
  sine 33
  standard 25
  Urban's 33
Curve-fitting 145, 149, 152, 154, 155
Curvilinear function 146

Degrees of freedom 6
Delaney clause 117
Density of organisms 52, 54
Design
  optimal 127, 128, 130–132
  sequential 143
  traditional 132
Deviation 16
  significant 17
Dichotomous data 124
Digamma function 114
Dilution 13, 52, 53, 58, 65
  geometric progression 65

Subject Index

Direct assays, *see* Assays
Dirichlet
   distribution 135
   prior 138, 140
   process 134
Distance function 74
Distribution
   algebraic 50
   angular 50, 142
   asymptotic 94, 126
   Behrens' 7
   beta 139, 144
   Cauchy 48, 142
   central t 7, 10, 11
   Dirichlet prior 134, 135, 138, 141
   extreme value 116
   fiducial 6
   logistic 102
   normal 5, 142
   omega 55
   posterior 135–139
   prior 135, 136
   tolerance 47–49, 51, 55, 58, 102, 133, 135, 140–143
   uniform 50, 53, 55, 142
   Weibull 123, 124
Dixon-Mood procedure, *see* Up and down procedure
Dose 14, 15, 24, 25, 28, 35, 38, 53, 54, 58, 60, 79
   allocation 9
   extreme 55
   high 58, 125
   level 33, 38, 40, 43–46, 50, 51, 56, 59, 64, 65, 69, 81, 103, 107, 108, 112, 117, 118, 122, 124, 125, 133, 143
   log dose 9, 24, 28, 29, 63, 86, 106, 112
   low 58, 113, 127
   response 1, 13, 55, 125, 128
   safe 117, 118, 121, 122
   spacing 106, 113, 132
Dose-metameter 14
Dragstadt-Behrens estimator 52
Dynamic programming 143, 144

ED_{50} 44, 58, 67, 69, 71, 85, 102, 105, 138
ED_{99} 57
Efficiency 48, 57, 112, 143

loss 112
Eigen vectors 111
Empirical probits 85
Estimate
   asymptotically normal 20
   best asymptotically normal 19, 43
   consistent 20, 120
   efficient 43
   maximum likelihood 61, 105, 119
   mean square error 109
   optimal 144
   prior 101
   ratio 4
   unbiased 5, 6
   weighted 7
Estimation
   interval 6
   moment 114
   points on the quantal response function 94
   precision 10
   simultaneous trial 25
   slope 24
   unbiased 114
   uniformly minimum variance 114
Estimator 109
   adaptive 141
   maximum likelihood 105
   nonparametric 134
Estrogen preparation 58
Euler's constant 53
Euler-MacLaurin formula 44, 47
Existence 125
Experimental error 152
Exponential approximation 114, 116
Exponential variable 115
Exposure time 122
Extrapolation 124, 125, 128
Extreme value distribution, *see* Distribution

Fatigue trials 95
Fiducial distribution, *see* Distribution
Fieller's theorem 5, 30
Finite experiment 45, 47
Fixation 28
Forced choice 69

Fractile 60
Freundlich's formula 29
Function
   curvilinear 146
   nuisance 118

Geometric progression 65
GLIM 69
Global maximization 126, 127
Goodness-of-fit 43, 56, 126

Haldane's discrepancy 41
Hazard rate 117, 118
Hellinger's distance 41
Heteroscedasticity 155
Homoscedastic 23
Horizontal distance 6
Hyperbola 150

Immunoassay 146, 148, 155
Inference, statistical 85
Infinite experiment 45, 47, 48
Inflection point 34, 47, 153
Information 47, 50, 51, 105, 137
   matrix 41, 49, 61, 83, 114
   prior 127, 133, 134
Insecticide 8, 12, 103
Insulin 145, 149
   immunoassay 148
Intercept 14, 25, 48, 82, 153
Internal consistency 37
Interpolation 131, 134, 150
   quadratic 100
Invariant 20
Isotope displacement 145
Iteration 83
   two-step 120
Iterative scheme or process 20, 23, 37, 40, 56, 98, 120, 138, 147
   Gauss-Newton 147

Kiefer-Wolfowitz process 96
Kolmogorov's inequality 97
Kullback-Leibler separator 41

Lagrange
   method 136
   polynomial 130
Langmuir's formula 29
LD_5 113
LD_{50} 45, 55, 58, 67, 80, 86, 103, 104, 108, 112

Subject Index

Least squares 36, 71, 121, 148, 151, 154
  nonlinear 130
  weighted 79, 91, 155
Ligand 154, 155
Likelihood
  equation 20, 36, 40, 67, 68, 82, 83
  function 20, 35, 56, 91, 93, 104, 114, 117, 119, 125–127, 136–138
Linear approximation 98
Logistic function 15, 29, 38, 44, 47, 48, 101, 102, 149, 153
  analysis 79, 92, 113
  difference 90
  effects 91
  parameters 90
  response curve 77, 81, 102, 111, 143, 153
  sum 92
  technique 154
Logit 35, 37, 55, 57, 60, 67, 69, 70, 71, 84
  approach or method 29, 33, 57, 59, 83, 113, 153
  estimates, optimal property 73
  modified 42, 43, 59
Loss function 133, 144
  quadratic 133, 139
Lung cancer 93
Luteinizing hormone 150

Markov chain 109–112
Matrix 150
  positive definite 78
  second derivatives 82
  transition probabilities 110
  Vandermonde 121
  variance-covariance 78, 122
Maximum likelihood 15, 19, 38, 49, 64, 67, 69, 81, 114
  estimates 57, 64, 102, 122, 126
  iterated 22, 70, 77
Means square 6, 7, 49, 85, 102, 149, 150
  error 106, 109, 110, 112, 142
Measures of association 89
Medians 102
Method or approach
  alternative 83
  cat 3
  comparison of various methods 83
  Davis 35
  dilution 101
  least squares 70
  linear regression 151
  maximum likelihood 20, 40, 43, 52, 78, 84, 85, 93, 104, 113, 131
  minimum chi-square 19, 40, 42, 43, 84, 85
  minimum logit chi-square 43, 84
  minimum modified chi-square 41
  moments 19
  parametric Bayes 140
  Pearl 35
  pseudo-Bayes 141
  scoring 56
Minimax 102
Minimum chi-square method, see Method or approach
Minimum modified chi-square, see Method or approach
Mode 20, 136, 137, 139
Models
  growth 34
  logistic 28, 85, 88, 105, 149
  logit 85, 105
  sigmoid 32
Monotone 13, 14, 109
Monotonicity 137
Monte Carlo studies 80, 85, 105, 109, 126
Mortality line 29
MUD procedure 112
Multinomial techniques 141
  formulas 141
Multiple sampling 111, 112

Natural mortality 40, 64, 67, 69, 85
Nematode species 85
Newton's method 96
Noniterative procedure 79
Nonlinearity 71
Nonparametric 45, 48, 100, 124
  Bayes approach 134, 140
  tests 126
Normal approximation 10
Normal equation 71
Normit 57

Objective function 120
Odds ratio 88, 92
Omega distribution, see Distribution
Optimal designs 59, 62, 131, 132
  dose level 62, 155
  stopping rule 96
Optimizing points 128, 130
Order statistics 93

P value 86
Parallelism 71, 87
Parametrize 60
Pathogenic bacteria 90, 92
Peaks 108–111
Pearson's chi-square 41
Penalty function 138
Phasing factor 109
Point of accumulation 120
Polygonals 153
Polynomials 153, 154
Pooled standard deviation 8
Posterior density 120, 136
  mean 144
  variance 144
Potency 14, 23, 30, 137, 146
  relative 8, 9, 24, 30, 71, 88
Precision of estimate 68, 98, 112, 150
Preparation
  standard 1, 8, 13, 14, 23, 29, 30, 31, 71, 88
  test 12–14, 23, 29–31, 71, 88
Prior density 20, 137
Probit 33, 38, 57, 69, 79, 83, 84, 86, 124, 125
  analysis 95
  estimate 73
    optimal property 73
  value 86
Propositions 16
Prospective study 89, 90
Provisional line 82
Provisional value 36, 64
Psychophysical experiment 69

Qualitative assays, see Assays
Quality control 151, 155
Quantal, see Response
Quantile 62, 95
  extreme 113
Quantit 56
  method 57

Subject Index

Radioimmunoassay 145, 152, 154
Random walk design 112
Range of doses 114
Ratio estimates 1, 5
RBAN (regular and best asymptotically normal) 74–77
Recursion 96, 100
Reed-Muench estimate 51, 52
Regression 6, 12, 13, 15, 16, 19, 22, 23, 25, 78, 85, 86, 102, 130
   coefficients 7, 87
   deviation from linear 16, 17, 19, 25, 87
   isotonic 136, 141
   linear model 23, 24, 151, 153, 154
   logistic 149
   metametric 15
   nonlinear 15
   robust models 155
   step-wise 26
   weighted 23, 25, 155
Relative potency, estimation 13
Reparametrize 61, 100, 136
Residual sum of squares 25
Response 1, 3, 10, 15, 20, 23, 24, 28, 29, 60, 103, 106, 131, 152
   angular 81
   dose 3
   graded 59
   logistic 78
   mean 2
   metameter 14, 21
   polychotomous quantal 57
   quantal 12, 22, 32, 48, 55, 59, 94, 100, 143
   quantitative 15, 23, 32
   type L 107
   type U 107
   working 22, 23
Retrospective study 89, 90
Robins-Monro
   estimator 102
   modification 102

procedure 101, 102
process 94, 95, 103
Robust 112, 113

S-shaped 28
Scatter 151
Second order efficiency 41
Sequential up and down method 112
Sheppard's correction 45
Skewness 42, 84
Simulation, *see* Monte Carlo studies
Slope 24, 25, 43, 82, 86, 111, 153
Spacing 44, 109
Spearman-Karber estimator, 44, 45, 48–51, 58, 112, 134, 141–143
Splines 153
Squared error 80, 135
Standard error 9
Statistic, Sukhatme's D 7
Statistical procedure 152
Sterling's formula 99
Stimulus 2
Stochastic approximation 95, 102
Stopping rules 108, 143
Strategy, optimal 144
Strong law of large numbers 20
Sufficient statistics 29, 43, 92
   minimal 36, 37
Sum of squares 88, 106
Survival 16
   data 18
Survivor functions 116
Susceptibility 28
Suspension 54

Terminal decision rule 143, 147
Test
   Bartlett's 18
   likelihood ratio 126
   nonparametric 126
   sensitivity 75
   simultaneous 25
Time to tumor 123, 126

Tolerance 11, 32, 33, 122, 133, 136, 139
Topology 133
Toxicity 60, 63
Transfer method 37
Transformation 17
   angular 17, 82, 83
   arc sine 19
   inverse logit 78
   linear for sigmoid curves 32
   log 34, 55
   logistic 89, 92, 93, 153
   metametric 15, 22, 29
   monotone 57
   scedasticity 22, 23
Transforms
   inverse logit 78
   logistic 89, 92, 93
   weighted 91
Transition probabilities 110
Turnbull-Bayes estimators 142, 143
Turning points 108, 109

Unimodal 46, 137
Uniqueness 125
Up and down procedure 95, 103, 107, 108, 112, 143
   modified 106
UTDR 111

Valleys 108–111
Variance 6, 8
   asymptotic 47–49, 54, 61, 100–102, 114, 130
   heterogeneity 17
   homogeneity 8, 9, 18, 19, 23
   homoscedasticity 17
   pooled 10
   of ratio 66
Vinyl chloride 131

Wallis product 100
Weight function 21, 37
Weighted mean 91, 92
Wesley's estimate 142